NETWORKING DEVELOPMENTS IN THE CAUCASUS REGION

NATO Science Series

A series presenting the results of scientific meetings supported under the NATO Science Programme.

The series is published by IOS Press and Kluwer Academic Publishers in conjunction with the NATO Scientific Affairs Division.

Sub-Series

I.	Life and Behavioural Sciences	IOS Press
II.	Mathematics, Physics and Chemistry	Kluwer Academic Publishers
III.	Computer and Systems Sciences	IOS Press
IV.	Earth and Environmental Sciences	Kluwer Academic Publishers
V.	Science and Technology Policy	IOS Press

The NATO Science Series continues the series of books published formerly as the NATO ASI Series.

The NATO Science Programme offers support for collaboration in civil science between scientists of countries of the Euro-Atlantic Partnership Council. The types of scientific meeting generally supported are "Advanced Study Institutes" and "Advanced Research Workshops", although other types of meeting are supported from time to time. The NATO Science Series collects together the results of these meetings. The meetings are co-organized by scientists from NATO countries and scientists from NATO's Partner countries — countries of the CIS and Central and Eastern Europe.

Advanced Study Institutes are high-level tutorial courses offering in-depth study of latest advances in a field.
Advanced Research Workshops are expert meetings aimed at critical assessment of a field, and identification of directions for future action.

As a consequence of the restructuring of the NATO Science Programme in 1999, the NATO Science Series has been re-organized and there are currently five sub-series as noted above. Please consult the following web sites for information on previous volumes published in the series, as well as details of earlier sub-series:

http://www.nato.int/science
http://www.wkap.nl
http://www.iospress.nl
http://www.wtv-books.de/nato_pco.htm

Series III: Computer and Systems Sciences - Vol. 181 ISSN: 1387-6694

Networking Developments in the Caucasus Region

Edited by

Jean-Paul Nadreau

Laval University, Quebec, Canada

and

Oliver B. Popov

Saints Cyril and Methodius University, Skopje, Macedonia

IOS Press

Ohmsha

Amsterdam • Berlin • Oxford • Tokyo • Washington, DC

Published in cooperation with NATO Scientific Affairs Division

Proceedings of the NATO Advanced Networking Workshop on
Networking Developments in the Caucasus Region
Tbilisi, Georgia
15–18 October 1999

ISBN 1 58603 179 1 (IOS Press)
ISBN 4 274 90455 5 C3041 (Ohmsha)
Library of Congress Control Number: 2001092449

Publisher
IOS Press
Nieuwe Hemweg 6B
1013 BG Amsterdam
Netherlands
fax: +31 20 620 3419
e-mail: order@iospress.nl

Distributor in the UK and Ireland
IOS Press/Lavis Marketing
73 Lime Walk
Headington
Oxford OX3 7AD
England
fax: +44 1865 75 0079

Distributor in the USA and Canada
IOS Press, Inc.
5795-G Burke Centre Parkway
Burke, VA 22015
USA
fax: +1 703 323 3668
e-mail: iosbooks@iospress.com

Distributor in Germany
IOS Press
Spandauer Strasse 2
D-10178 Berlin
Germany
fax: +49 30 242 3113

Distributor in Japan
Ohmsha, Ltd.
3-1 Kanda Nishiki-cho
Chiyoda-ku, Tokyo 101
Japan
fax: +81 3 3233 2426

PRINTED IN THE NETHERLANDS

CONTENTS

List of Participants

AKHUNDZADE Adil

IATP
1 Sharifzade, Baku, AZERBAIJAN
E-mail: adil@iatp.baku.az

AMIRBEKYAN Vahan

Internet Program, OSI — ARMENIA
E-mail: vahan@osi.am

ATOEV Aso

OSI
TAJIKISTAN

BELYAEVA Alexandra

Head of Communication Reasearch Center,
Institute of Psychology of the Russian Academy of Sciences,
Maximova st. 8-3, Moscow 123098, RUSSIA
E-mail: abelyaeva@vega.ru

BEREZHNEV Sergei

Radio-msu, Moscow State University, RUSSIA
E-mail: sfb@radio-msu.net

BONAC Marco

ARNES,
Jamova 39, 1000, Ljubljana, SLOVENIA
E-mail: bonac@arnes.si

CHACHKHUNASHVILI Michael

Executive Director, Open Society — Georgia Foundation,
Tbilisi, GEORGIA

CHILADZE Badri

Deputy Director, High Energy Physics Institute,

Tbilisi, GEORGIA

CHKHENKELI Nino

USIS — Tbilisi, GEORGIA
E-mail: ninochk@access.sanet.ge

COTTRELL Roger

Stanford Linear Accelerator Center,
Mail Stop 97, Stanford Linear Accelerator Center, P.O. Box 4349,
Stanford, CA 94309, USA
E-mail: cottrell@slac.stanford.edu

DAVIS Michael

Central Asian Development Agency, Internet Project Director,
TADJIKISTAN

FRASHERI Neki

Institute of Informatics and Applied Mathematics (INIMA),
Tirana, ALBANIA
E-mail: nfra@inam.al

FRESE Hans

Deutsches Elektronen Synchrotron, Hamburg, GERMANY
E-mail: hans.frese@desy.de

GAJEWSKI Jacek

CEENet Secretary General,
Pasteura St. 7, Warsaw 02-093, POLAND
E-mail: gajewski@ceenet.org

GARDNER Keith Leroy

Deputy Assistant Secretary General for Scientific and Environmental Affairs, NATO
Bd. Leopold III, B-1110 Brussels, BELGIUM
E-mail: science.netw@hq.nato.int

GARIBASHVILI Devi

Department of information Georgian Academy of Sciences,
Tbilisi, GEORGIA

GIORGADZE Helen

NATO National Office of Georgia,
Tbilisi, GEORGIA

GRUNTORAD Jan

CESNET, Association of Legal Entities,
Zikova 4, 16000, Praha 6, CZECH REPUBLIC
E-mail: jan.gruntorad@cesnet.cz

HESS Juergen

Open Society Institute, Soros Foundations,
888 7th Ave, New York 10106, USA
E-mail: jhess@sorosny.org

ISSAKHANLY Hamlet

AZERBAIJAN

JANZ Robert

Computing Centre, University of Groningen,
9700 AV Groningen, P.O. Box 800, The Netherlands
E-mail: r.f.janz@rc.rug.nl

KAFFENBERGER Walter

Programme Director, NATO Science Division,
Bd. Leopold III, B-1110 Brussels, BELGIUM
E-mail: w.kaffenberger@hq.nato.int

KAKUBERI George

Deputy Minister, Georgian Ministry of Telecommunication and Post,
Tbilisi, GEORGIA

KARUMIDZE Sandro

Internet Center for a Wide Open World,
Open Society — Georgia Foundation,
Tbilisi, GEORGIA

KIKNADZE Levan

Deputy Director, Institute of Computational Mathematics,
Tbilisi, GEORGIA

KIRTAVA Zviad

National Information and Learning Center (NILC), Director

KONTRIDZE Vakhtang

IREX Internet Access and Training Program,
Internet Program Coordinator,
2 Chavchavadze Avenue, Tbilisi, GEORGIA
E-mail: vkontridze@access.sanet.ge

KRAAV Mihkel

Estonian Educational and Research Network Tahetorn,
EE2400, Tartu, ESTONIA
E-mail: mihkel@eenet.ee

KURAKAEV Magomed

Ventindustria Company, Head of Board of Directors,
Ventcommunication (VENTCOM), General Director,
Mayakovskogo St. 3, Grozny 364007, RUSSIA
E-mail: mamed@grozny.su, ventcom@grozny.su

KUYKENDALL Jacob

GTE, USA
E-mail: jacob.kuykendall@gte.net

KVATADZE Ramaz

Deputy Director, High Energy Physics Institute,
Tbilisi, GEORGIA

MAFTER Ilya

OSI, New York, USA
E-mail: imafter@inch.com

MKRTCHYAN Ruben

Director of Yerevan Physics Institute, Br. Alikhanian St. 2, Yerevan, 375036, ARMENIA
E-mail: mrl@arm.r.am

MKRTUMYAN Igor

President of Armenian Internet User Group, American University of Armenia,
Bagramyan St. 40, Yerevan, 375019, ARMENIA
E-mail: imkrtoum@aua.am

MULKIJANYAN Grigori

Egrisi, Ltd Systems Administrator,
#16 Jikia St., Tbilisi, GEORGIA
E-mail: egrisi@caucasus.net

NADREAU Jean-Paul

Associate Vice-Rector, Research, Office of the Vice-Rector Laval University,
Quebec City, Qc, Canada, G1K 7P4, CANADA
E-mail: nadreau@vrr.ulaval.ca

NANOBASHVILI George

Internetional Telecommunication and Information Center,
Technical Manager,
#2, 26 May Square, Tbilisi, GEORGIA
E-mail: gia@itic.org.ge

NAVOTNY Vadim

Physical-Technical Institute,
Mavlyanova St. 2, 700084 Tashkent, UZBEKISTAN
E-mail: navotny@uzci.net

PETROSYAN Artur

Armenian Academy of Sciences,
Liebknecht St. 14, Yerevan, 375010, ARMENIA
E-mail: hovik@undp.am

PITSKHELAURI Tamaz

Head of Department, High Energy Physics Institute,
Tbilisi, GEORGIA

POPOV Oliver B.

Sts. Cyril and Methodius University,
Krste Misirkov bb, MK-1000 Skopje, MACEDONIA
E-mail: oliver@marnet.mk

PROKHORENKO Eugene

YerPhI, Ministry of Post and Telecommunications, Nalbandyan St.28,
Yerevan, 375010, ARMENIA
E-mail: drostom@acc.am

PSHENICHNIKOV Michael

Internet Program, KYRGIZSTAN
E-mail: michael@soros.kg

PURVES Alex

Loreal Orian, UK
E-mail: apurves@loralorion.com

RASTL Peter

Vienna University Computer Center,
Universitaetsstrasse 7, A-1010 Wien, AUSTRIA
E-mail: peter.rastl.univie.ac.at

SADOWSKY George

Director of Academic Computing Facility,
251 Mercer St., New York 10012-1185, USA
E-mail: george.sadowsky@nyu.edu

SANDU Nicolai

ROMANIA
E-mail: nsandu@cisco.com

SARTANIA Vakhtang

Rector of Tbilisi State Pedagogical University,
Tbilisi, GEORGIA

SASIEDZKI Piotr

Silesian Institute of Technology Computer Center, POLAND

SENA Nadezhda

President, Open World Foundation for a Sustainable Future,
P.O. Box 19252, Boulder, CO 80308, USA
E-mail: nadia@mystico.colorado.edu

SIMONYAN Armen

Armentel, ARMENIA

SOBOLEV Valeri

UZBECNET
UZBEKISTAN

STUBBS David

Executive Director, OSI-Azerbaidzan, AZERBAIJAN
E-mail: dstubs@osi-az.org

SULABERIDZE Zviad

Sientific Institute Techinformi, Tbilisi, GEORGIA
E-mail: zvi@mmc.net.ge

TAVKHELIDZE Albert

President of Georgian Academy of Sciences,
Tbilisi, GEORGIA

TAVKHELIDZE Ilia

Deputy Director of Institute of Applied Mathematics,
Tbilisi, GEORGIA

TONEYAN Albert

Head of Computer Center of Yerevan Physics Institute,
Yerevan Physics Institute, Br. Alikhanian St. 2, Yerevan, 375036, ARMENIA
E-mail: albert@r.am

TSKHOVREBOV Alan

VILLEMOES Peter

NORDUnet
Agern Alle 3, DK-2970 Hoersholm, DENMARK
E-mail: peter.villemoes@nordu.net

VILLEMS Anne

EE2400 Tartu, ESTONIA
E-mail: anne@ut.ee

VITKOVKSY Vladimir

Chief of SAO, RAS Informatics Department,
SAO RAS, Nizhnyi Arkhyz, 347147 KChR, RUSSIA
E-mail: vvv@sao.ru

VRGA Vesna

Croatian Academic and Research Network,
J. Marohnica bb, 10000 Zagreb, CROATIA
E-mail: vesna.vrga@carnet.hr

WENZEL Zita

Ph.D. USC/ISI, 4676 Admiralty Way, Marina del Rey, CA 90292, USA
E-mail: zita@isi.edu

YILMAZ Tugrul

TUBITAK
Tunus Cad. 80, 06100 Kavaklidere, Ankara, TURKEY
E-mail: tugrul-yilmaz@metu.edu.tr

ZEYNALOV Vugar

OSI-Azerbaijan,
39 Azadlig Ave. Apt. 52, Baku, AZERBAIJAN
E-mail: vzeynalov@osi-az.org

Organization

NATO-country Co-Director

J.-P. Nadreau
Laval University, Quebec, Canada

Partner-country Co-Director

A. Tavkhelidze
Georgian Academy of Sciences, Tbilisi, Georgia

International Organizing and Programme Committee

A. Belyaeva
Institute of Psychology, Moscow, Russia

M. Chachkhunashvili
Open Society — Georgia Foundation, Tbilisi, Georgia

H. Frese
Deutsches Elektronen Synchrotron, Hamburg, Germany

J. Gajewski
CEENet, Warsaw, Poland

R. Kvatadze
High Energy Physics Institute, Tbilisi, Georgia

R. Mkrtchyan
Yerevan Physics Institute, Yerevan, Armenia

J.-P. Nadreau
Laval University, Quebec, Canada

O.B. Popov
University St Cyril and Methodius, Skopje, Macedonia

Local Organizing Committee

B. Chiladze
High Energy Physics Institute

D. Garibashvili
Presidium of the Georgian Academy of Sciences

S. Karumidze
Open Society — Georgia Foundation

L. Kiknadze
Institute of Computational Mathematics

T. Pitskhelauri
High Energy Physics Institute

I. Tavkhelidze
Institute of Applied Mathematics

Jean-Paul Nadreau

Summary and Acknowledgements

The NATO Computer Networking Programme supports two types of Advanced Networking Workshops (ANW): the Training workshops and the Policy workshops. The aim of a Training workshop is the transfer of technical know-how to network administrators, whereas in the case of a Policy workshop, it is to increase the level of harmony among the various network policies at national and international levels. These workshops provide a forum for users and policy-makers and those responsible for the scientific infrastructure to acquire a better understanding of the need for co-ordination in the planning and development of research networks.

The workshop held in Tbilisi in October 1999 was of the Policy Workshop type, with a regional focus on the Caucasus region. This workshop can be counted as one of those Policy Workshops that have, in the past, been initiated by NATO in countries where a regional strategy in networking was deemed necessary. It was a natural continuation of the Policy workshops that were held in Budapest in 1993, in Moscow in 1994, in Almaty in 1995 and in Tartu in 1997, when I was the Computer Networking Programme Director. It was clearly a pleasure for me to be the NATO co-director of this Workshop. I would like at this time to thank the community for its trust and confidence.

The Context

Due to their belonging to the High Energy Physics community, the first links to the region took place through the Yerevan Physics Institute (YerPhI) and through the High Energy Physics Institute of Tbilisi Sate University (HEPI TSU). These connections were established between 1994 and 1996 with the support of NATO and INTAS.

Apart from these flagship projects that played a crucial role in the development of the Internet in the region, the coverage of this part of the world remains minimal. Even if local connectivity has dramatically improved lately (for example, in Yerevan) thanks to the generalization of the use of radio-modems, providing connectivity to remote academic and research institutions is still a challenge. In Georgia, a need to connect the cities with research communities such as Kutaisi, Batumi, Poti and Rustavi has been clearly identified. From the Armenian viewpoint, connectivity to Iran on one side and to the Black Sea neighbouring states through Georgia on the other side was considered as the expected outcome of the fibre optic line under construction.

However, these developments face the type of problems already identified in other regions such as the Balkans and the Baltic States. The problems may vary from country to country, but they were accurately listed in the Tartu Declaration (1997) as being:

- A lack of awareness among political decision makers;
- Inadequate support for network initiatives;
- A lack of co-ordination among different national actors;
- Some fragmentation of the efforts;
- Inadequate training facilities;
- A shortage of personnel, material and infrastructure resources;
- High tariffs for national and international circuits;
- A possible misunderstanding of the role of individual research, academic and educational networks by Public Network Operators (PNOs); and
- Continuing monopolies.

The need for a concerted strategy

One of the challenges of the workshop was to go beyond the physical (the mountains) and the political (the large number of small states and territories) barriers in the region.

Consequently, the prime objectives of the workshop were:

- To give the opportunity to network managers and policy makers to meet in person and exchange views on the present networking situation in their respective regions;
- To help them define the needs of the region and the possible areas of co-operation;
- To provide them with the proper information and expertise for planning projects in the near future.

A total of 66 participants from 25 countries attended the Workshop.

The outcomes

The workshop and the follow-up meeting that took place in April 2000, also in Tbilisi, provided the participants with a better understanding of the regional situation and of the possible synergies and led to a series of recommendations on several fronts such as infrastructure, training and regional organization.

We all can be pleased with the major outcome of the follow-up meeting that led, on April 28, 2000, to the signing of a Memorandum of Understanding between Networking Organizations from Armenia, Azerbaijan, and Georgia that stated:

The undersigned representatives of the following networking organizations do hereby declare the will to work together to realize common networking goals in the Caucasus Region.

The main goal is to foster the stability, growth, and internal sustainability of the Internet infrastructure for the socio-economic development of the Caucasus Region. Desired outcomes of this international networking include increased political stability, free exchange of views, and rebuilding of links between the scientific and educational communities of these states of the former Soviet Union.

Increased communication between these three countries will enable scientists and educators to work in close collaboration, to exchange ideas and results, and to work together on projects of common interest (e.g. sharing of online catalogs between national libraries and centralized training in library science, seismic data collection and exchange, Internet infrastructure redundancy in case of earthquakes, creation and maintenance of common usage databases such as flora and fauna and cultural heritage of the Caucasus Region). Additionally, large-scale collaboration with scientists in other countries sometimes requires regional mirrors of central databases (e.g. genetics, high-energy physics, etc.).

International traffic will be optimized by direct routing between Caucasus Region countries. In addition, mirroring the redundancy structure of the Internet, regional collaboration on Internet infrastructure will allow back-up services and alternate routing in case of failure or overloading in one of the countries.

The development of a Caucasus regional network, based on existing national infrastructures, will provide an easy access point for integration with other European and global networks, which will become increasingly important for growth and international communication.

The coordination of Caucasus countries' networks will encourage information technology standardization and increase compatibility and exchange of information. Lastly, the formation of a group of country networks will allow bulk purchases and to negotiate deeper discounts with vendors.

To realize these goals, and to coordinate and leverage the use of resources and funds, we express the desire to create a single management organization (working name Caucasian Academic, Research, and Educational Networking Association — CARENA) The organization would comprise one representative organization from each country, to be determined. In the case of Georgia, the single representative would be GRENA (Georgian Research and Educational Networking Association).

Acknowledgements

This workshop and the follow-up meeting could have not taken place without the remarkable commitment of a large number of people. My sincere thanks go to the members of both the local and international Organizing Committees. Prof. A. Tavkhelidze, President of the Georgian Academy of Sciences and co-director of the Workshop, and I are particularly indebted to key players in the organization such as Jacek Gajewski, Oliver Popov, Ramaz Kvatadze, and Sandro Karumidze. My warmest thanks to them.

I would like to conclude by expressing my gratitude to the NATO Scientific Affairs Division and its highly devoted staff, who placed their programmes at the service of the scientific community. I commend the Computer Networking Programme and the members of its Advisory Panel for the vision they have expressed in managing the programme.

Keith Gardner

Foreword

The Internet is changing the world we know in fundamental ways. Businesses, governments, academic institutions, the news media, the entertainment industry, and communications are all being deeply effected by this technology. But no sector of human activity is being more deeply impacted than science. The Internet was born of the need of scientists to collaborate effectively with distant colleagues. The Internet initiative is has now producing a truly new environment for human activity, and one which continues to evolve at a breathtaking pace. The Internet environment is insensitive to distance. It provides virtually instant access to people and information resources literally all over the world. It is also an information-based infrastructure with very little physical inertia. It is not dependent on bricks, concrete, factories, roads or consumable resources. Rather, it resembles a virtual city with information structures as "buildings", with standards as the "building codes" and with bandwidth as the "streets and highways".

NATO's Computer Networking activity, under the NATO Science Programme, is aimed at assisting the Partner nations in bringing their scientists more fully into this new environment. The provision of Internet access to a scientist results in orders of magnitude increases in capability. This results from the ready access to colleagues, publications, data and models from the global scientific community.

This Advanced Networking Workshop was the fifth such event in the series of policy workshops sponsored by the NATO Computer Networking Panel addressing regional issues. Previous Workshops were held in Budapest in 1993, in Moscow in 1994, in Almaty in 1995 and in Tartu in 1997. The Tbilisi Workshop has already yielded valuable results, which promise to help the nations in the Caucasus region to benefit more completely from the Internet revolution.

I was personally pleased to participate in the Workshop, and on behalf of NATO I extend an expression of gratitude to the organizers and to the Georgian hosts of this very successful event.

Keith Gardner
Brussels

Albert Tavkhelidze

Welcome to Georgia

Dear Colleagues,
Ladies and Gentlemen,

Allow me on behalf of the Georgian scientists to welcome you — participants of the international workshop and to thank our foreign guests for coming to Tbilisi.

Specialists from 25 countries will be exchanging ideas on networking problems. Georgia is represented at the Workshop by the institutes of the Academy of Sciences and Universities.

This Workshop is organized jointly by NATO and the Georgian Academy of Sciences. We are grateful to NATO for the financial support that has made it possible to hold this Workshop.

The Internet-based network incorporating the institutes of the Georgian Academy of Sciences and State Universities is successfully functioning in Georgia. The network was developed thanks to grants from NATO and the European Union. I should particularly note the great importance of the network to our scientists by assisting them in establishing contacts with their colleagues abroad. These contacts are essential for the promotion of science. We hope that the workshop will help in the further development of the Internet network in Georgia and generally in the Caucasus with the kind support of NATO.

Once again I would like to thank all the organizations and their representatives who have assisted in the creation and functioning of the scientific and education network in Georgia.

I wish for the success of the Workshop and for our future cooperation.

Jean-Paul Nadreau

NATO in Tbilisi — October 1999

NATO ANW — Tbilisi, October 1999

- Welcome to Tbilisi.
- 5th NATO Policy workshop:
  - Budapest, Hungary, 1993,
  - Moscow, Russia, 1994,
  - Almaty, Kazakhstan, 1995,
  - Tartu, Estonia, 1997.

Tartu Declaration

- The lack of awareness among political makers;
- The inadequate support for network initiatives;
- The lack of co-ordination among different national actors;
- Some fragmentation of the efforts;
- Inadequate training facilities;
- Shortage of personnel, material and infrastructure resources;
- High tariffs for national and international circuits;
- A possible misunderstanding of the role of individual research, academic and educational networks by Public Networks Operations (PNOs);
- The continuing monopolies.

Objective 1

- To give opportunity to network managers and policy makers to meet physically and exchange views on the present networking situation in their respective regions.

Objective 2

- To help network managers and policy makers define the needs of the region and the possible areas of co-operation.

Objective 3

- To provide network managers and policy makers with the proper information and expertise for planning projects in the near future.

Expected outcomes

- The workshop should provide the participants with a better understanding of the regional situation and of the possible synergies;
- This should lead to a series of recommendations in various fields.

In terms of structure

- One of the workshop's objectives is to help the proponents bring their projects to a greater state of maturity so that they can secure the required funding;
- Several other projects should also come out of the workshop. This workshop should be the ideal forum for co-operative projects.

In terms of training

- The experience of CEENet in training should be useful. It will help the participants define their own needs and devise proper training programmes for the network administrators and the users;
- This should lead to well-structured training programmes and to various proposals, be they to NATO or to other agencies, for Training Workshops in the region.

In terms of regional organization

- The development of the Internet has been directly linked to the establishment of national networking associations. Academic networks have been instrumental in the cohesion of this development;
- Managing a network requires skills and experience. A minimal structure is needed to make things happen: securing funding and organizing training requires it;
- This workshop should provide us with basic principles from which we can organize our own environment;
- This is also a perfect opportunity to set up alliances and achieve better connectivity.

This workshop will be a success if we are ready to:

- Listen
- Share
- Learn from each other, and
- Be creative

We are all here in ***TBILISI*** *. . .*

- ***T***o
- ***B***uild
- ***I***nterconnectivity and
- ***L***earn
- ***I***n
- ***S***haring
- ***I***nternet

Keith Gardner

Welcome on behalf of the NATO Science Division

Mr. Chairman, Distinguished Guests,
Ladies and Gentlemen,

I'm honored to welcome you to this advanced networking workshop on behalf of NATO.

This year marks the 50th anniversary of the creation of the NATO Alliance in 1949. Many things have been happening in the world over those 50 years. As a part of the celebration for the 50th anniversary at the NATO Summit in Washington, we were asked in our division to come up with a very short statement of what we do. That's not easy. The NATO Science Program includes many areas and many activities. But we were able to came up with seven words that encapsulate what the program does, and those seven words are: "Bringing scientists together for progress and peace". This really does describe the most important thing which we do in the program. Bringing scientists and people together — this is the core, the heart of NATO Science.

Some years ago, I had a personal experience which has convinced me of the importance of this concept. I worked in the United States as a researcher for the US Navy for 15 years prior to coming to NATO. At one point, our group was involved with doing automatic ship classification using high-resolution radars. This is not something that everybody does. There were few groups, even in the United States, who were working in this area. In the course of our work, we came into contact with a research team in the United Kingdom which was doing almost exactly the same thing. We were able to get a data exchange agreement set up between the United States and the United Kingdom in this area. I traveled to UK to meet with this team, and to set up a network with people who were doing the same sort of work that we were. This contact was extremely stimulating, because the UK team was doing the same things in different ways, with different equipment. And by comparing notes both teams were able to have insights that we could not have gained otherwise. It is very interesting to note that those insights were easier to achieve with people from another country than they were with people from my own country, because we weren't competing with them for funding. So, very often crossing international borders allows you to be more candid than you can be with people from your own country. From this experience and others, I have a personal conviction that bringing scientists together can be very stimulating.

There was an initiative in the US after the Second World War which I find enlightening. There was the time in the 1950's when the situation in Europe was rather desperate for science. There wasn't much funding for travel, and the US arranged for the European scientists to be able to travel to the US on the military aircraft for the cost of one US dollar. So, they paid their dollar, got on the aircraft and travelled across the Atlantic to sit down and work with US scientists. This was a very successful program. I'd like to note that we're now doing essentially the same thing — we're getting scientists to travel so they can work with their counterparts in other countries. The cost today is more than one US dollar per trip. But the real cost is not that great when one takes into account the benefits and the impact, considering all of the resources throughout the nations which are devoted to scientific activities. By getting scientists together we exchange information, and exchange knowledge which leads to progress, which is one of our seven words. Also by getting together, by exchanging information, by working together and socializing, we create a trusting relationship. This trust within the scientific community leads to trust between nations, and this in turn leads to stability and then to peace, which is another of our seven words.

I really believe that today's workshop is getting a very important community working together in a way which can contribute to stability and peace.

This workshop is a promising one in two important ways. First of all we enhance personal face-to-face contacts, which you are directly engaged in during this workshop. You are all going to exchange with each other your views, your backgrounds, and your understanding of the needs that you have and possible approaches. This process brings comprehension of what is possible, of what is needed to your countries and regions, and that's very important. Face-to-face contacts are crucial in this respect. They bring understanding of each other, and stimulate new thinking.

The second important factor is that we're dealing with the Internet, which represents a true revolution. This is something that world has never seen before. In fact, we are dealing with a new world, which is a "virtual" world. What does the term "virtual" mean to me? It means that we are not tied down by all of the existing infrastructure of the real world. We are not tied down by physical systems. I'm going to use a word that I'm not sure the interpreters know, which is "leapfrog". In American it means that you're jumping over steps in a process to get to a more advanced state, without going through all of the previous states. And this is what joining the Internet allows us to do — we leapfrog technology, we do not have to go through the building of the infrastructure of buildings and roads, or circuits and cables and wires — you're going directly to capabilities which are not dependent on previous development. Once you have the connection and the bandwidth you can get the access to all of the processes that are available.

I wish you all success in your exchanges and discussions of the coming days, and extend to the Georgian authorities NATO's thanks for accepting to host this important workshop.

Giorgi Kakuberi

Telecommunication Development in Georgia

Dear Mr. Chairman,
Dear Guests,

I would like to welcome participants and organizers of the advanced workshop held by the Georgian Academy of Sciences and NATO.

Organization of such workshops, and participation in them is topical for Georgia — for the country that has recently regained its independence and which is undergoing constructive process.

The impact of the large-scale changes associated with information dissemination is of the same significance as the impact of other known events that occurred in the past. Actually they can be considered a real revolution in the information and communication field that changed the world. Hence, the 20th century has prepared the material basis for the new century: as it is known, the 21st century will be the century of globalization and personalization in telecommunications.

Implementation of new technologies in parallel with the existing telecommunication environment will ensure the reality of a new, unified telecommunication environment for the countries of the South Caucasus, Central and Minor Asia, and of the Black Sea region.

Georgia is a small country located at the crossroads of Europe and Asia. Taking into consideration the above geopolitical factor, revival of the ancient "Silk Road" through accomplishment of a new "Silk Road" and TRACECA (Transport Corridor — Europe–Caucasus–Asia) projects is under the way thanks to the by initiative and personal efforts of Mr. Eduard Shevardnadze, the President of Georgia. Information and technology societies are already formed in certain countries of the contemporary world. Developed countries of Europe are on the way of promotion of such societies. Bearing in mind the principles of the information society, one cannot consider development of information and communication technologies only in the national context.

Considering the geopolitical factor of Georgia, the development of information and communication technologies must be envisaged as the laying of fiber-optical backbone lines along TRACECA, the new "Silk Road", as well as the creation of an independent telecommunication corridor, laying fibber-optical backbone lines along the Trans-Asia, Europe and South Caucasus railway road; and the incorporation in future of the Poti–Varna Poti–Rize fiber-optical lines into the general network that will give them strategic and economic significance.

It is envisioned that through the above backbone line will pass the telecommunication traffic of the following countries located along the Europe–Asian corridor: Transcaucasian and Middle Asian countries, China, as well as Bulgaria, Ukraine, Turkey and other countries of Western Europe. Many countries and companies show interest in the above projects.

Many countries of the world support the new "Silk Road". The "Act of the Silk Road Strategy" passed by the US Congress in 1999 from this viewpoint can be considered as an event of great importance. The Act reflects issues of the strategy of democratization, economic integration, and political and economic stabilization in the South Caucasus, Central Asia and neighbouring regions. The importance of telecommunication development is especially stressed in this comprehensive document, and this implies support of the US Government in reconstruction and further development of the existing information and communication infrastructure. The strategy developed in this document, the revival of the ancient "Silk Road" will strengthen stability in the South Caucasus, and will promote a process of regional integration. In turn, the latter will pre-determine the economic and political originality of a country.

Taking into consideration the above, we think that integration of information and communication technologies into the European and the world telecommunication system will become a decisive factor for strengthening security of our country, for its economic development, cooperation and development.

Development of the global information infrastructure for many countries has become the political priority that determines the future of the country. As Mr. Al Gore has figuratively mentioned, the global information system "definitively unites mankind" because we all are closely bound up with each other in the information era. Communication and information technologies are the eye of economics, the necessary factor for its development.

Today we can say with confidence that we are undergoing a renaissance of telecommunication and information technologies. The tendency toward the large-scale implementation and accessibility of new technologies is obvious in the 20th century.

As we know, the mass implementation of radio broadcasts took several decades (some 40 years), that of television — a dozen years, and that of the Internet — just 2–4 years.

There comes a time when the power of a state is defined not only by industrial or military potential, but in the first place by timely incorporation into the world global information space; creation, accessibility and use of information resources.

Coverage of our country with information and communication technologies will give incomparable possibilities for dissemination of extensive data in different fields of our life. Privatization of the communication field is under way in Georgia and this makes the implementation of new technologies the most important question.

Integration of the telecommunications of Georgia into the worldwide telecommunication network is associated not only with technical and technological issues, but also with the solution of the wide spectrum of economic, information and technology problems in general.

Recently an Enactment of the President of Georgia "About activities for development of the information and communication technologies" was signed with the purpose of solving this significant problem. This can be considered a new stage of forming the information-technological society. The information-technological society must become the common feature of our country in the 21st century. The Coordination Council for the Promotion of Information and Communication Technologies of Georgia was set up on the basis of the above Enactment. The President of Georgia chairs the council. The council consists of prominent scientists, specialists in various areas, engineers, and ministers.

The Council has to prepare the long-term presidential programme for information and communication technological development including ideology and practice for forming of information and communication society, the necessary prerequisites for its development. The Programme will reflect the role of the information society, and will address issues such as: electronic commerce, telemedicine, customs, insurance, functioning of teleinformation and other services, technical, technological and methodological issues of distance education.

Taking into consideration the benefits of computer networks and technologies implementation, the Ministry of Communication and Post demonopolized and liberalized this field in Georgia and simplified the license-issuing procedure.

It is well known that implementation of information technologies requires significant investments. We do hope that our friends and partners will once again grant their support, giving Georgia the chance to rank with countries that have a high level of information technological development.

Thank you for your kind attention.

Networking Developments in the Caucasus Region
J.-P. Nadreau and O.B. Popov (Eds.)
IOS Press, 2001

George Sadowsky

The Internet: Reflections and Visions

Introduction

My name is George Sadowsky, and my professional background is in mathematics, economics and computing. For 13 years I worked for the United Nations doing technology transfer of information technology, and after that I became involved in academic computing in the USA, in particular most recently managing networks and network services at New York University. I have worked for the Internet Society as a volunteer since 1992, mainly in the organization and execution of the network training workshops for students from developing countries.

The purpose of the first speech of a conference is to motivate discussion on a number of topics, to set a stage, to be provocative and to leave the audience challenged and (hopefully) awake at the end of the talk. So for this reason I will be provocative intentionally, I will express a number of personal opinions, and some of you, I am sure will disagree with me, perhaps violently, but there certainly will be time for discussion afterwards.

Networking is a very broad topic and it is difficult to know exactly what to talk about, but I would like to concentrate on several areas. I would like to make as a theme the need to ask the right questions. This is a topic which is taught in the universities often — it is no good to have the right answer to the wrong question, it does not help. It is no good to do a wrong thing efficiently, it does not help, and it does not give you a solution you want.

I am going to talk about the following areas:

1. The information revolution and a notion of ‘Internet time’, which has already been mentioned by several speakers;

2. Technology and scalability — where are we and what can we expect in a near and a long future;

3. The notion of Internet governance;

4. The tension between growth and equity, which is sometimes called ‘The Digital Divide’;

5. Cultural and social issues — I will give some comments on the cultural and social issues, raised by the growth of the Internet;

6. Issues for developing regions — some issues that affect developing countries and developing regions of the world.

The information revolution and 'Internet time'

The information revolution has been building up for a long time.

We often think that the information revolution is very new. In fact there has been a revolution in the way we think about information for thousands of years.

We depend very much upon insights which seem trivial to us now, but in fact were major accomplishments of their times.

For example, on the computing side:

- Concept of numbers, algebra, geometry;
- Concept of negative numbers also (imaginary numbers);
- Mechanical single calculations (1700+);
- Pre-defined calculations (1800+):

  – Babbage's engines,
  – Hollerith's machines — invention of the punch card;

- Electronic calculation (1943+);
- John Von Neumann's breakthrough of the stored program;
- Ambiguous memory, storing both programs and data (making possible assemblers and compilers);
- Commercialization of computers (1950+);
- Emergence of a mass market for computing (1975+).

The commercialization of computers is only 50 years old and in those last 50 years a mass market for computing has emerged only in the last 20–25 years. So, if you look at the whole history of information and information technology you can see that there is an enormous backdrop, which has made possible the development of computing, but the take-off point — at least for widespread use of computing technology — occurred only about 20 years ago.

Then, on the communications side we have the same development:

- Speech (5000 B.C.?);
- Writing (3000 B.C.?);
- Printing (mass distribution of writing), Gutenberg (1500+);
- Telegraph (1844);
- Telephone (1890?);
- First man-computer communication (1938?);
- Telex (first widespread electronic mail (1950?));
- Digital communication (1955+?);
- Initial time-sharing (1960+);
- Initial packet sharing (1969).

We are getting close to the Internet now . . .

- NCP (Network Control Protocol (1970?));

- ARPANET (1971);
- ISO/OSI data communications model (1979?);
- TCP/IP Implementation (1983 — full implementation);
- NSFNet (1985);
- Commercialization of the Internet (1990+).

We have the development of speech — thousands of years ago, then of writing using appropriate character set. The notion of mass distribution of writing, which was initiated by Gutenberg through his invention of the printing press, only occurred about 500 years ago.

So, if we look at the progress of information technology, we can observe that it has been advancing for centuries. But it seems to us that it is taking off only now and we are living in what is often called 'Internet time'. And one of the problems I think we have, is that the Internet and the technologies that support the Internet are proceeding at Internet time but our brains are not. We still live in human time and it is often a very frustrating experience to try to keep up with the pace of change.

The industrial revolution as a comparison

The industrial revolution is often used as a comparison to the information revolution. If one examines the industrial revolution, one finds that it also occurred over many hundreds of years. However, in contrast, the information revolution very much depends upon SOFTWARE, the stuff of thought — it is much easier to change a computer program than it is to change the design of a steam engine. So, the time it takes to reshape the world of information, as opposed to the mechanical world, is very short and leads to this sense of wonder that things are proceeding so quickly.

The period from 1750 to 1930 (i.e., from the introduction of the steam engine to the use of mass production assembly lines for automobiles) was a period of profound changes, but the rate of technical progress was relatively slow — at least it seems so in retrospect — and depended upon manufacture of large physical goods.

Nevertheless, if we measure the information revolution starting from the telegraph, then the length of the revolution is not dissimilar.

The information revolution, as we know it, has been characterized by a rapid rate of progress and is measured in different ways. Often we measure it in terms of a productivity–price ratio or a performance–price ratio.

This is a ratio which by most measures has been increasing by about 25–30 percent a year for the last 40 years. Over 40 years the compound growth rate has been extraordinary and almost defies comprehension.

There is an American writer, Garrison Keillor, who has invented a community called 'Lake Wobegon', and this is such a nice and good community that all the children are above average. Now you all know that is not possible. I think that in the Internet community it is possible for the average person to feel below average. And for good reason — there is not enough time to learn everything that is happening. What that says is that the average person does not know as much as we think he might, and that is very frustrating for us because this is an interesting field and those of us who are professionals want to keep up with new developments and have great difficulty doing so.

Technology and scalability

For the last 40 years we have achieved a rate of progress of 25–30 percent per year. What will then happen next year? in 10 years? in 20 years? The technology is shifting dramatically. There are some 'laws' of computing and networking development worth discussing that bear on the future.

Some have stood the test of time, others have not.

- Grosch's law came first, but now is dead. In the late 1940's and early 1950's Herb Grosch stated that 'the value of a computer is proportional to the square of its cost and size'. Now this was in fact true of the early 1st, 2nd and 3rd generation of computers, where there were significant economies of scale, and when computers cost a lot more than people, but it is not true today where economies of individual use and control dominate any computational economies of scale that may continue to exist.
- Metcalfe's law seems well entrenched — 'the value of a network is proportional to the square of the number of people who are on it'.

In other words, the more people who are on the Internet, i.e., the more people who are connected to the community — the more important and useful the net is.

Moore's law says that the number of transistors on a chip doubles every 18 months, and since it takes about the same cost to build the transistor you have twice as much power every 18 months. But how long will Moore's law continue to be valid?

For the last 30 years there have been people who have said that this rate of progress cannot go on. I remember a discussion in 1987 with a physicist who worked for one of our top security agencies in the United States, who said that 'at the one micron level that is the smallest width of the possible path for electrons, so it looks like we will have reached the theoretical limit of chip manufacture'.

That was said by a person who had a Ph.D. in physics, and was guiding future chip development.

However, 12 years later we are now building chips with 0.25 micron widths or maybe even 0.18 micron widths. One week before this workshop there was an article in the New York Times which said: 'physicists at Intel now announce the fundamental barriers to chip technology'. Will it happen? — we do not know. Other people in the article said — 'whenever there has been a fundamental barrier, people have found a way around it'.

There are some fundamental barriers, but we are not anywhere near them.

The scalability of the technology will not stand in the way of increase in price-performance for quite some time.

Here are some interesting questions:

- How many IP addresses will there be 20 years from now, and what will they be attached to?
- Cellular telephones;
- Household appliances?
- Note: 802.11 wireless is rapidly being deployed;

- Note: blue-tooth chips are 1+ years from now at $5–10;
- Note: BACNet is being widely deployed.

BACNet is a new protocol used for a network of automatic controllers, which will be IP-based next year. Now I operate a large network at the New York University, and recently I found that, unknown to me, people building the new buildings at my University are installing almost 20 sensing devices in every room of each of these buildings, so that each new building on my campus has at the moment the possibility of having tens of thousands of sensors or devices, which could be providing information on some kind of a network. Sometime, somewhere, and it will not take long before those devices become active devices and have the possibility of controlling other things. So, the danger is that there will be millions and millions of devices which we do not recognize now as network devices but which may be on the network at some future time.

Thus the answer to our question is that billions and billions of IP addresses will be needed.

What will be the geographic spread of the Internet?

Almost all countries have now a point of presence, and there is rapid growth of the Internet in many countries, slower growth in some others. Some of you have probably seen the maps showing the penetration of the Internet which Larry Landweber produces. Those maps are not very useful now because almost all countries have some Internet connectivity. For example, the Republic of Mali connected to the Internet in 1996. Now in many places in Mali, the Internet is a local phone call away. There is even a telecentre in Tombouctou.

What contribution can we expect from Internet-2?

Some of you have heard about Internet-2 as a new vision of the coming Internet, with the notion of differential qualities of service (QoS) and high bandwidth as the immediate payoffs.

Although Internet-2 is still at an experimental level, nevertheless it is spreading. Jan Gruntorad, who is here at this workshop from CESNET, was with me at the Internet-2 annual meeting in Seattle at the beginning of this week. I remember 1994, five years ago, when Jan and I and about 50 other people worked for several months to put the first E1 Internet circuit into Prague. This was for the Internet Society Conference and as I recall, there was not enough demand to sustain it afterwards. Last week Jan signed a memorandum of understanding with the Internet-2 organization to put an E3 circuit, a 34 megabit terrestrial link to Prague directly from North America. He also told me that he was in the process of establishing a 1 gigabit link between Prague and Brno in the Czech Republic.

This is major change. The countries of Eastern Europe and the Caucasus are somewhere in the middle of this accelerating revolution, certainly not the slowest but maybe not the fastest.

Governance issues

I would like to talk for a few minutes about governance issues, because this is the subject where Internet time and human time collide. For several reasons this is perhaps in the longer run one of the most difficult issues that we will leave to face.

One reason is that we are chasing a moving target; this thing that we are trying to govern or manage is continuing to change in front of us and grow, so that the rules that we would have adopted for governing yesterday's Internet may no longer apply.

A second reason is that we are not politicians. We tend to be technical or come from a technical background, and what we are trying to do is to develop a political system that is different from any political system achieved till today. Clearly, to develop a system for governing this uncontrollable thing that changes as we watch it is not an easy task.

Here is what I think the right question is: What should be the governance model for the Internet in the long run?

In particular I want to comment on two issues:

- What is the role of the state in Internet governance?
- Is it possible to realize democracy in cyberspace?

Our attempts so far have been mixed. The Internet was born in North America, and it has been sheltered and nurtured by the United States government in general quite well. The transition to privatizing the net has been handled with mixed success, but a disastrous decision in 1993 gave to Network Solutions Inc. a monopoly in domain name space registration.

We are still suffering from the major negative effects of this decision. The development and the formation of ICANN, which some regard as the proper organization to begin to implement an administrative structure for the Internet, has been torturous, and the difficulties surrounding its birth tells us something about what it takes to implement democracy in cyberspace.

We also confuse management of the Internet with governance of the Internet. The state will take care of governance. The state has interests to prevent illegal activity, to protect the right of confidentiality, to regulate commerce, and to ensure national security. Nothing that Internauts do is going to make the state forget that, and the state is going to intervene as it feels necessary to ensure that those aspects of behavior are controlled. But different states will intervene in different ways, and that is part of the problem.

International organizations also have an interest in the Internet. The International Telecommunications Union (ITU) now understands the importance of the Internet, after a slow start. The Organization for Economic Cooperation and Development (OECD) is interested in electronic commerce. The European Union (EU) is very concerned about the confidentiality of individual information. The World Intellectual Property Organization (WIPO) is involved with regard to the trademark issue.

I claim that much of the success of the Internet so far has been due to a bottom-up methodology of development, and that we would not have had the same development if a central authority had been in charge. The State is not very good at bottom-up development. So, when we talk about the management of the Internet and a place for its governance, then probably the national government is the last place one should go to.

So where does this leave us? The technical standards for the Internet have been largely governed by an organization called the Internet Engineering Task Force (IETF). It is a meritocracy — it does not matter who you represent, and it does not matter what country you come from. What matters is that you have good ideas and that you express them well and you interact well and respect the ideas of others. Can we scale this to a nation-wide or world process? It is not clear. The IETF has about 2500 people who attend its meetings, but the world contains now 6 billion people.

Although far from perfect, the ICANN is perhaps the best organization for driving the management of the Internet at this time. And the structure that is proposed for it has various standing bodies concerned with addresses, names, protocols, etc. There was a large argument for the last year about how should individuals, how should people, users be represented in ICANN? How should the governance process work? Should everyone in the world have a vote? Should only users have a vote? What is the process if we want to continue this democratic participatory model of governance? What is the process by which we get there? And so the issue is — what is the Internet community and what does it mean to have democracy in cyberspace? What are the rules of behavior?

Well, on the Internet at the moment the rules of behavior are not very polite. Cyberspace is big. We do not know the individuals who write the messages we read, we are almost always far away from others who read what we write, and conversely. And what seems to happen is that instead of a convergence of opinion we very often observe a polarization of opinion. This behavior, which becomes less and less courteous, is called flaming. And we do it, and it generally does not really help solve any of our problems. So, the question is — what does this imply for the success of self-governance on the Internet?

In the real world there is a park in London — Hyde Park, and it is famous for people expressing their opinions. Anyone can go to Hyde Park, bring a box to stand on, get on the box and start talking. And people come and listen to the speakers; they gather around to either listen or disagree. Now everyone who speaks in Hyde Park is equally loud, and that is in some sense what happens on the Internet. We have learned that everybody can be equally loud on the Internet, and that the people who write the most or are able to get to the microphone the most are seen as expressing stronger opinions than those who do not. What does this mean? How will we get ourselves out of this problem, in which the noisy people hold more sway that the quiet ones? — 'You may go to Hyde Park to express your opinion, but Parliament does not meet in Hyde Park!'

Growth and equity: the digital divide

I would like to talk a little bit about the growth and equity issue. This is perhaps one of the most difficult issues that we have and it has local, national and international dimensions. There is often a tension between growth and equity. Equity refers to degree of relative equality in some dimension across a set of people. Different political systems have different opinions about the degree of equity that should be provided for its citizens and how it should be provided. For example, in primitive capitalism, growth was important and equity did not matter. In some forms of socialism and communism — equity mattered a great deal but growth was less important. Socialism insures equality, and as one of my Hungarian colleagues commented in 1976, it insures that everybody is equally miserable.

That reflects the tension. But the question is not how do we eliminate inequality, although it is a very politically correct question to ask. We give recognition to the fact that some people are ahead, some are behind but it will always be that way. The question we should be asking is: what is the proper balance between these objectives and how can we harness the tension between them to achieve the balance and the growth that we want?

My sense is that in the short run the successful model for getting this distribution of resources to move forward is to empower the private sector, i.e., the people who are greedy, the people who want to make money, and let them advance in that direction as far as they can, in the direction of more progress. Then it is the function of government and other organizations, e.g., NATO, the Internet Society, and non-governmental organizations to push the trailing end of the curve to make sure that the distance between the frontier and the people at the end is not too great. Of course, that is a value judgement. What is not too great? Clearly, people will have different opinions about that.

The longer-team strategy for improving access has to rely upon the education and the educational system to spread knowledge and the benefits of information technology. Locally we have the possibility of government making strategic investments in technology, in telecenters in developing countries, as well as in cyber-cafés, where the private market cannot provide them.

There is an interesting finding that has been discovered in this area. There was an experiment in Boulder, Colorado, in the United States, in which the Internet was extended to low-cost public housing. The intent of the experiment was to enable Internet access for people who perhaps did not have jobs and were not aware of benefits they could receive from using the Internet. However, what happened was not what was expected — the Internet was used that way, but not very much. Instead, the people in the housing development built their own web site on a server and used it to explain issues of poverty and unemployment and the relevant characteristics of their life to the greater community. There was a reverse effect and I think that was a consequence that paid off in a big way.

Government intervention can help — there are a number of countries in which the Internet is a local call away — I have already mentioned Mali, which is going in this direction. Such a system is also in place in Spain as well as in Malaysia, and perhaps also in some other countries ... Where you are a local call away from the Internet, even if the call may be metered, you have a greater degree of access to the Internet.

The international situation is what interests us here. There are a number of programs andinternational organizations which can help in terms of lessening the digital divide. Not all of them are big as they should be, and not all of them achieve their objectives as well as they should. There are indirect programs such as those executed by the United Nations Secretariat that offer sectoral assistance. The ITU is helpful in arguing for the elimination of tariffs, decentralization and privatization of telecommunications carriers. NATO helps by giving advanced workshops and by funding people to learn about the Internet. The Internet Society and non-governmental organizations (NGOs) help by providing training and assistance of various kinds. We all wish was more money for this, but in fact the money delegated to the amelioration of the digital divide is the result of a political process in which nations and individuals — such as George Soros — decide what they are willing to provide as a budget to support this cause. I should have mentioned George Soros earlier, since he is one of the major benefactors of this region, as well as of the Internet Society workshops.

Let me talk now about cultural and social issues.

Here are two questions: Is the Internet a threat to national cultures? And the second question is — What is the culture of the Internet and does it add to other cultures? How does it interact with other cultures?

Let us address the second question first. The Internet initially was characterized by openness, sharing, helpfulness, a small community of people working together, believing generally in North American values, because that is where it grew up. The current Internet has shifted toward commercialization, advertising, misuse: spam, chain letters, scams, mischievous hacking, etc. But if it were not for the commercialization of the Internet, without a massive amount of investment both of money and information into the Internet — today's Internet would not exist.

And what we have in the background is that the spirit of cooperation and mutual help still exists in many places, e.g., Internet Society, NATO workshops, and the Leland Program in Africa. We also have the spirit of cooperation that continues to exist in many of the personal relationships that we have as Internauts among ourselves. So that still exists and it is important for us to keep that alive in the face of the continuing trend of Internet-1 toward the commercial markets. That trend gives us other benefits. It gives us for example the ability to access amazon.com and have a book delivered to us quickly no matter where we are in the world. That is a major benefit.

The first question has raised issues in several countries. Is the Internet destroying national culture? I think that the answer is — not really. The merging of cultures has been promoted for thousands of years from the time that the first caravans and ships travelled between two countries. International travel has become very commonplace and continues to infuse cultures with elements of other cultures. More recently, the jet plane has intensified that process. Radio and television have brought into everybody's lives a notion of what is happening in other parts of the world. Satellite TV has brought CNN and other stations to everyone on the planet who is able to receive their broadcasts, and that is a lot of people.

I say here with some embarrassment that the US film industry, Hollywood in particular, has managed to create a film culture representing itself as the culture that has propagated throughout the world. The film culture is not America, of course, but in the process of propagating itself, my opinion is that it has done more harm than good, except possibly to those who have financially benefited from selling it in countries outside the United States.

What does the Internet add to these already existing communications processes between countries and cultures? It adds inexpensive, almost instantaneous two-way communication. And that is dangerous if a nation wants to maintain cultural homogeneity and cultural identity. The Internet also implicitly promotes English as the second world language. One could argue, however, that English occupied that place already as the common-denominator language of business between nations.

However, the Internet also supports specific national and linguistic identities — slowly, computer software and web servers are beginning to adopt UNICODE, which includes 16-bit representations of all currently used alphabets in the world. UNICODE makes possible the easy development of information in a local alphabet and in a local language in such a way that it coexists with other languages using other alphabets. The Internet also promotes a process which has been called a reverse diaspora. The concept of diaspora is that of the emigration of people from their home country to many different places all over the world. It has happened in many countries. The Internet allows that process to reverse on a virtual basis so that, for example, French Canadians in Winnipeg (an English-speaking

Canadian city) can communicate and share information with French Canadians in Quebec much more easily than they could if they had to depend solely upon mail or telephone or newspaper.

Cultural clashes are difficult to handle because they challenge deeply-rooted beliefs and feelings. There are three areas in which such clashes generally become very serious. The first is the treatment of sexual matters. The second area is criticism of government and authority, and the third area is freedom of, or from, religion and religious expression. And all those combinations are represented in today's world. So if the information from country A is retrieved in country B, governments may or may not take offense, and may or may not try to block the transmission of that information. What will happen, what will this be like in 20 years? This is a problem that is not going to be solved in Internet time — it is going to be solved in human time. There are number of countries that restrict access to content although they can only do it imperfectly, they cannot do it completely.

In Prague in 1994, when George Soros was the plenary speaker at INET '94, he asked the right question from the stage. He said: 'How many independent Internet connections do we need to go into the former Soviet Union and link it with the outside world so that never again can a government stop information from flowing in and out of this region?' That is not precisely the right question, but it is surely the right idea.

Personally, we are faced with the dilemma with regard to these restrictions on an everyday basis. As an individual, working with governments, I often do not criticize such restrictions, I do not stand on principle, saying — all information must be allowed to come into the country. I do not feel the necessity to do this because I have faith when once you open the door to Internet access, it cannot be closed. Vint Cerf said many times, 'the Internet never retreats'. On the other hand, for organizations such as the Internet Society, we have to stand on that principle and we have to say that content restrictions are unacceptable in today's world. If a government chooses to impose that there is nothing we can do, for the Internet Society is very small compared to the size and power of a typical national government. But on the other hand, information will bypass the restrictions. Sometimes it is just a matter of making a telephone call to a neighbouring country to get the information you need.

Let us talk a little bit about issues concerning developing countries.

There are lots of opportunities for developing countries to use the Internet. You all know that. The most important by far is education, because it is in the long run the best solution for investing in human capabilities and catching up with the developed world. Especially in this area of the world, where even though economic conditions are not as good as elsewhere, the educational system has been good and education and science have historically been highly valued. Good science has come from this region. So it is really important to continue the investment in education and to make the content resources of the world, not only in science, available to as many people in the region as possible to establish those collaborations and ensure that the educational process is as strong as possible.

There are other reasons — commercial reasons for countries here to tell the world about themselves. They have the capability through the Internet for a global audience to explain who they are, what they do and to reduce the barrier for foreigners to learn about the benefits of working with you and visiting you, either on business or as tourists.

There is another concept, an economic concept called opportunity cost that is important in this context. The opportunity cost of a decision is measured in terms of the cost of opportunities foregone.

For developing countries, the cost of not keeping up with the world is measured in terms of becoming more non-competitive and more dependent upon external forces and events. As the information revolution continues to develop, information increasingly translates into economic power, with access to educational opportunities, market information and the ability to enter markets. Developing countries need to have these options, and the ability to exploit the Internet in multiple ways is a key to obtaining and maintaining such options.

The Internet is now everyone's business. Developing countries have more to gain than developed ones, because they lack the alternatives and substitutes often available in developed environments. Furthermore, it is in the interests of both the developing and developed worlds to ensure a connected world, for economic growth, for the peaceful solution of international issues, and for maximizing the human potential available to all of us in the world. This conference, sponsored by NATO, is one of the vehicles that will help lead to such a more connected and a more prosperous world for everyone.

Thank you very much.

Tamaz Pitskhelauri

An Overview of the HEPI Network

The Networking communication center of the High Energy Physics Institute (HEPI) of Tbilisi State University (TSU) provides Internet connectivity for the research institutes of the Georgian Academy of Sciences (GAS) and Georgian Universities. Currently, HEPI connects (and provides basic Internet access) to 31 research and educational organizations. The Institute of Computing Mathematics of GAS also has Internet access via HEPI, with a large networking system — AcNet and the VIAMNet networking system of the Vekua Institute of Applied Mathematics (VIAM) of TSU. VIAMNet alsoprovides Internet access to all the departments of Tbilisi State University.

The satellite link from HEPI to DESY operates, since September 8, 1999, at a speed of 128/64 kbps in the transmission/reception mode accordingly and saturation of the channel is considerably less than at the previous speed of 64/64 kbps.

Within the Tbilisi area, three types of connections are used for the remote access to HEPI:

- Leased lines
- Dial-up
- Wireless

In the near future, based on fiber optics, a link will operate between HEPI and VIAM (for adistance of approx. 1 km) with a speed of 10 Mbps.

Since the beginning of this year, the radio-networking system operates within the Tbilisi area. The grant awarded by NATO provided the necessary support for this project which started in 1998. Thanks to the project, HEPI received very fast and reliable connections among the main research institutions in Tbilisi. Indeed, the goal of the project has been to create wireless links for the Internet access and data exchange between research centers and educational institutions in Tbilisi.

The Radio-networking system has two main centers — the University high building (UHB) and the high tower near the Institute of Computing Mathematics upon the roof of which are installed main repeaters omni-directional antennas of HEPI and ICM. These antenna serves as a star type network center. The connections between the main nodes, as well as between HEPI and UHB and tower and ICM, are realized by the directional antennas. The scientific and educational organizations are divided into two basic groups according to the location that enables optical visibility.

There are nine main computing centers connected by the radio-network:

1. High Energy Physics Institute of TSU
2. Institute of Computing Mathematics of GAS
3. Institute of Physics of GAS
4. Tbilisi State University (main building)
5. Presidium of the Georgian Academy of Sciences
6. Razmadze Institute of Mathematics
7. Wide Open World (Internet center of OSGF)
8. Department of Sciences and Technology
9. Medical Diagnostic Center

Advantages of the Radio Network

- Easy to install
- Easy to extend
- Independent link
- Speed 2 Mbps

Problems of the HEPI Network

- Satellite terminal 'Kedr'
- Link to DESY
- Routing at DFN

New Equipment for Upgrading the Satellite Station

- Satellite terminal CST-5000 EFData — 40 w
- LNA — 35 K
- SDM — 300 A

The equipment was made possible by a NATO grant for 1999.

Networking Developments in the Caucasus Region
J.-P. Nadreau and O.B. Popov (Eds.)
IOS Press, 2001

Levan Kiknadze

The Status and Plans of the Development of the Georgian Academ-Network

Exactly five years ago the Georgian Academy of Sciences received a grant ($150,000) from NATO for the development of telecommunication facilities to avoid the information vacuum and to integrate in to the world. Dr. J.P. Nadreau, Dr. H. Frese as well A. Beliaeva and S. Beliaev have taken an active part in this great event and we are very grateful to them. A little later, INTAS awarded another grant of $100 000 for HEPI at TSU. These two grants and the appropriate projects have been merged to realize a united computer R&E network for the Scientific Community and Higher Education system. Two Network Operation Centres (NOC) — one in HEPI at TSU and a second one in the Institute of Computational Mathematics (ICM) of the Georgian Academy of Sciences have been established.

Thanks to the grants ICM received the following equipment: Ultra-Sun SRARC, Router Cisco 2511, modems — US Robotics, a high speed modem Nokia BB512 kbps (for connection HEPI-ICM through telephone cable), Ethernet's components, UPS — 6,5 kva, Diesel–Generator — 2 units.

The main duty of our NOC is the access to the Internet, primarily for the scientific institutes of the Academy of Sciences and another research institutes also. Out of the 66 academic institutes (see: http://www.acnet.ge) at this moment, Academ-Network includes 45 institutes and 9 institutes that are connected to HEPI network due to their close location.

In addition, our responsibility is to provide the 12 regional branch institutes (located in 5 towns of our country) with access to the Internet. These intentions will be realized with the National Research Education Network (NREN) project which is the subject of a special presentation at this ANW.

Before we state our current situation I must underline the gigantic problem which absolutely affects our activities at both the international and local levels. Three years ago, the rate of our satellite channel was quite acceptable, but today, when local ISPs have 512–1200 kbps, our connectivity is too slow (even for a service free of charge). Although we are very grateful for the recent increase in the rate up to 128 (after 3 years' stagnation), we hope that in the near future there will be more support from God and well-wishing persons.

It was also a very appropriate time to receive the second grant from NATO (one year ago) for the realization of the project concerning the building of the wireless network. Thanks to this, we achieved quite reliable connections to the satellite station — HEPI. It was a very important event which included several institutes. However, because of the restricted amount of radio-equipment (16 units), the difficulties of most of the users remain valid. I must mention that this network was constructed under the guidance of HEPI.

The actual and ordinary problems

After that I will address other difficulties:

I think it is well known that the global problem of FSU cities is the poor state of telephone cables. That is why one of the permanent tasks is to overcome the "problem of last mile" for our end-users who could not get more or less normal connectivity. It is clear that communication based upon our city's telephone cable-net is almost hopeless if we do not use the new (for us) technology — HDSL (High bit rate Digital Subscriber Line). As we know, this technology is very popular in the western countries and in Moscow also. This gives us hope for using xDLS even in our conditions. Investigation of this problem is being carried out by the specialists of the Faculty of Telecommunication of Technical University, who are also analyzing the possibilities of application to the peripheral towns.

Unfortunately, there are zones in the academ-campus where the phone cables are in a terrible state. Not one of the nine institutes can use phone cables for a normal connectivity to the NOC (although the NOC is located close to the campus). The optimum solution was to construct an ordinary campus LAN that connects all nine institutes. The area of the campus is less then one square km. The names of the institutes are:

1. A. Razmadze Mathematical Institute
2. A. Janelidze Geological Institute
3. K. Zavriev Institute of Constructing Mechanics
4. M. Nodia Institute of Geophysics
5. National Seismic Defence Centre
6. V. Bagrationi Institute of Geography
7. K. Kekelidze Institute of Manuscripts
8. Central Academic Library
9. Geophysics Barochamber
10. Institute of Computational Mathematics (NOC)

Each one of these institutes is equipped with the necessary facilities for an access to the Internet. The Mathematical Institute excels, having the radio-bridge, received thanks to the INTAS grant for publication of the international *Georgian Mathematical Journal*.

There are two versions of the draft project. The first one proposed building the Ethernet LAN based on fiber-optics technology. The fibre cable can be laid down in the existing underground channels.

The second version considers the usage of two radio modems (ARLAN 640-2400), UTP cables and 9 hubs. The two radio modems provide connectivity to 8 buildings, which are united through UTP cables (distances between the buildings are 60–100 m). The ninth building is the Barochamber with highest tower where functions an omni antenna according to the last project. This antenna is a point of visibility for eighteen institutes.

Both versions are easy to implement and the initial cost is 11,000–12,000 USD for each of the alternatives.

There is one more urgent problem concerning the NOC. For its elementary functioning it is necessary to increase the capacity of the communication equipment of our NOC. At this moment, we have only 9 dial-up and 5 leased lines. This is certainly far from a normal set up for 45 institutes with about 3000 scientists, even if we take into account 10% of users. So, it would be advisable to install one Cisco 3600, 16 USR courier modems and 8 new phone numbers. For the connection to the telephone exchange there is a separate cable.

A few words about the activity of our team

The duties of our team (composed of eleven persons) include a facility to help users access and search the Internet. For this service there is a proxy server. In the near future there will be a mirror server, too. We hope that these servers will, to certain extent, help our users to obtein the desired information more quickly and reduce the overload of the link.

One of the activities of team includes the support and development of the Web site of the Georgian Academy of Sciences: www.acnet.ge. On this site, there is information about 64 institutes, scientific departments and members of the Academy, and general data about our country. There are lots of links to Russian Web sites for users who do not know foreign languages.

Our NOC is based at the Institute of Comp. Mathematics where for a long time there has been a team of programmers specializing the creation of data bases. At this moment, they are studying the enormous and unique information present in many academic institutions and collected for decades. The data bases will be dedicated to the national wealth in history, archeology, rare museum artifacts, ecology, seismology, traditional medicine, fauna, flora and the natural resources. Through the support of our team these data bases will be put on our web-server.

I am sure that this server will become a very popular and attractive one for Western scientific and commercial users. Then, the problem of increasing the rate of the satellite channel will become also their problem.

Networking Developments in the Caucasus Region
J.-P. Nadreau and O.B. Popov (Eds.)
IOS Press, 2001

Ruben Mkrtchyan and Albert Toneyan

Developing Networks for Science and Education in Armenia

This is a brief history of Yerevan Physics Institute's participation in the development of networking in Armenia. Networks connected to YerPhI, with nodes in scientific-educational organizations, will be described. In particular, informational and financial flows of these networks will be described, with some suggestions for future developments.

1. Development of Internet Connection through Yerevan Physics Institute

1993 — dial-up connection to Moscow
1994 — 64 kbps/sec satellite line to DESY

- 128/64 kbps/sec (1996)
- 196/128 kbps/sec (1999)

Second channel:

YerPhI-Armentel-Teleglobe/MCI 128/128 (1999)

Third channel:

Asymmetric, based on DirecPC technology, 400 kb/sec (1998)
1998 — A compression method on satellite channel
1998 — A local internet registry, a full member of RIPE
1999 — BGP and dynamic routing

2. Development of networks

1995 — YerPhI–NAS collaboration: radionetwork in Yerevan (640 kbps/sec, 5 nodes) (Cylink)
1996–1997 — Radionetwork in and around Yerevan (2 Mbps/sec, 11 nodes, Aironet)
1998–1999 — Radionetwork in and around Yerevan (11 Mbps/sec, >60 nodes)

Direct connections to all ISP's inside Armenia

3. Maps of the networks

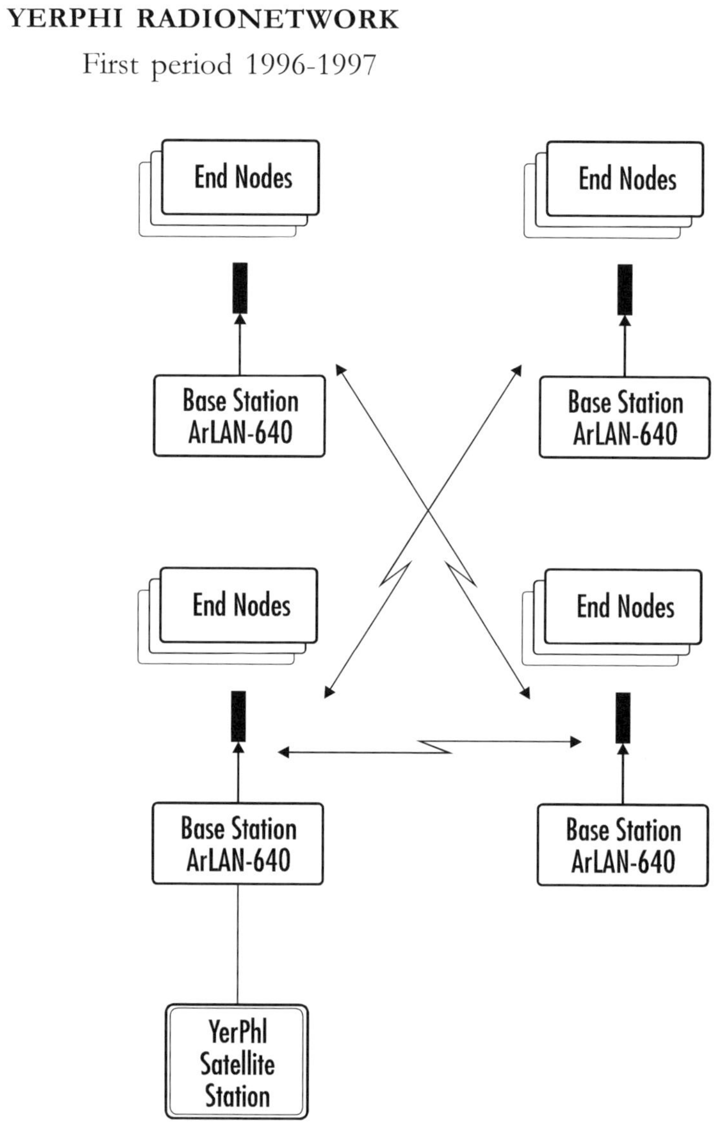

YERPHI RADIONETWORK

Second period 1997-1998

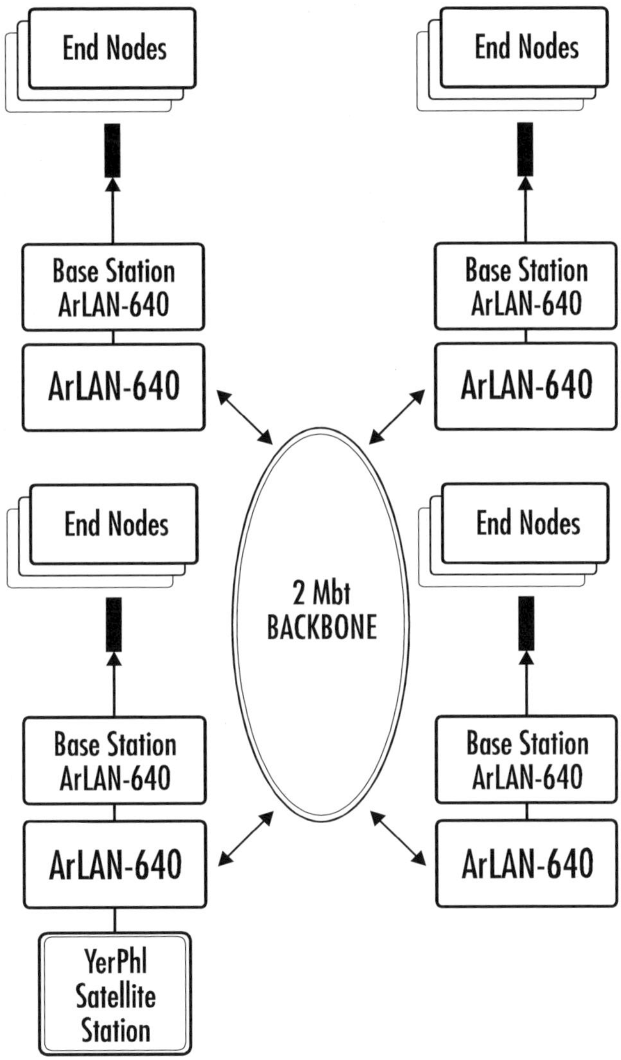

YERPHI RADIONETWORK

Third period 1999

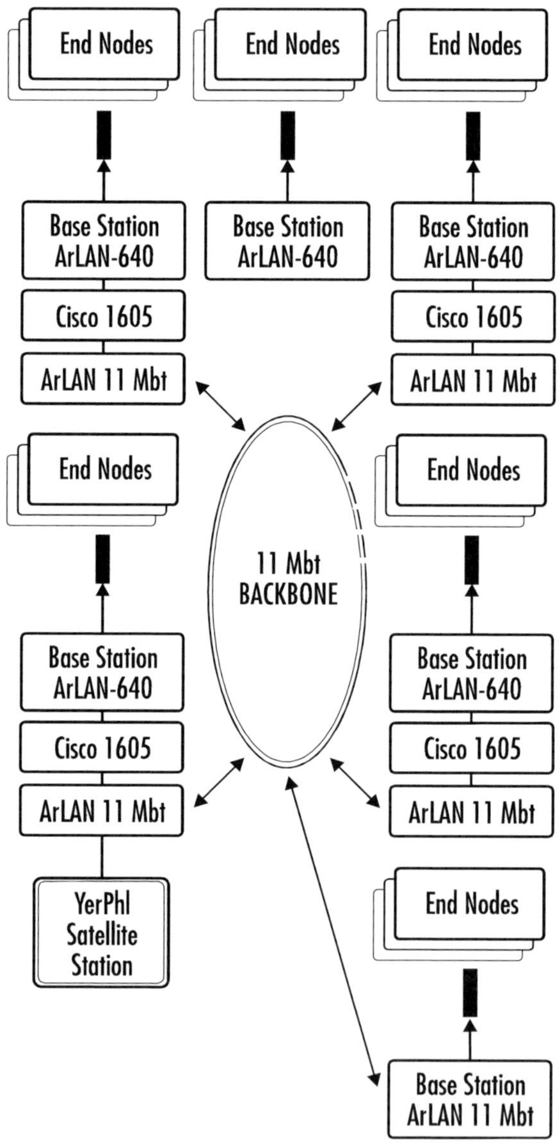

4. Users

Currently, 14 class C networks are divided in 46 subnets with 450 hosts and globally routed IP. More than 50 hosts are proxy servers, with an average of 15 workstations behind each. Thus more than 1200 computers are connected to the Internet via YerPhI.

Educational and scientific organizations (35) are the main users of our network. They provide more than 75% of traffic to Internet. In addition, more than 20 governmental, nonprofit and other organizations are also connected.

The traffic for some of these organizations is presented below (in %):

	am	bus	life	misc	net	Uni	unkn.
AI	38	24	22	8	2	2	4
MSRC	19	4	7	17	16	4	33
YSU	25	3	3	3	7	57	2
CIT	80	2	4	2	3	0	9
Diagn	59	3	8	8	8	7	7

AI — Architectural Institute
MSRC — Molecule Structure Research Center
YSU — Yerevan State University
CIT — Center of Informational Technologies
Diagn — Medical Diagnostical Center

5. Organization of the service of the network

Running the Internet connection and the network: service of satellite lines and backbone were provided by one organization (YerPhI), sites in the institutes — by common efforts of local organization and YerPhI.

Financial flows.

From the beginning, the price introduced was to cover the operating expenditures. Prices were and still are a few times lower than that of any commercial ISPs. The costs are connected to capacity in Mb and some quotas. The objectives are 1) — to provide a cheap Internet; 2) — to cover operating costs; and 3) — to provide a quality of the Internet connection, by imposing some restrictions on an unlimited use of external lines, which can cause problems for other users.

The problem of operating of such big networks is important. The need for flexible and fast management is not always possible to meet in a big scientific organization. It is reasonable to hire (or create) a special organization, devoted to providing a network service, which otherwise would remains the property of scientific institutes and universities.

6. Needs for future development

- Faster connection (prices, funding).
- Clear ether for the development and operating of radionetworks.

- More reliable telephone lines or some slower and chipper wireless devices for use at home, at least for e-mail only.
- Contacts with experienced ISP providers for consultations on the organization of large networks, control and billing systems.

Networking Developments in the Caucasus Region
J.-P. Nadreau and O.B. Popov (Eds.)
IOS Press, 2001

Vugar A. Zeynalov

Experience and Perspectives of the Information Networks Development in Azerbaijan

Background

History

In 1996, Chevron and the Open Society Institute Internet Program collaborated to provide connectivity to the Academy of Sciences in Azerbaijan. While Chevron provided connectivity, the OSI-IP provided a number of terminals for a public access center. Internet activity in the Academy of Science has proceeded slowly. After repeated attempts to develop an Internet project with the Academy of Sciences, OSI-Azerbaijan moved forward to create its own Internet facility.

The OSI — Azerbaijan Internet Program efforts in Azerbaijan started in *November 1997*. After the evaluation study in mid-November 1997, it was decided to establish an open-access Internet site under the auspices of OSI-Azerbaijan. On April 15, 1998, OSI-Azerbaijan Public Internet Center offered its services to wider audience for the first time.

Introduction

Currently, Baku has two projects acting independently, but sharing connectivity, to provide Internet and e-mail access to the general public: the OSI-Azerbaijan Internet Program and the Internet Access & Training Project (IATP), implemented by IREX with USIA funding. OSI and IATP are cooperating to establish a data network comprising academic and research institutions and public sites. We hope that the NATO Science Program will become an OSI/IATP partner soon, providing a NATO Networking Infrastructure Grant.

Information data network

General description

Established in the fall of 1998, the OSI/IATP network became the starting point for the implementation of projects to enhance Internet infrastructure, both in Baku and in the regions of Azerbaijan. By providing low-cost, unlimited Internet access *to qualified organisations*

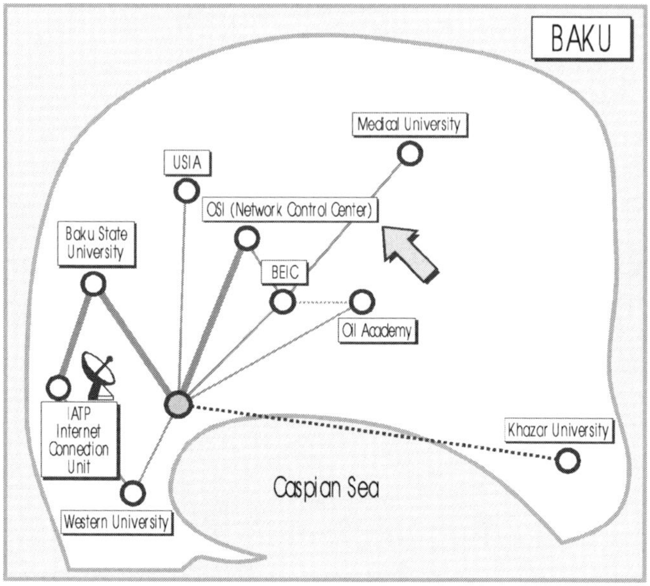

OSI Network Control Center

and individuals, the network has formed the foundation of a national academic information network.

The Central Communication Unit: All centers connected to the network use a common communication facility, allowing OSI and IATP to effectively share a common satellite Internet connection, which ensures the dynamic balancing of the traffic load and ease of network management.

The Network Control Center: The center provides the following services and resources for other network facilities:

NCC Services and Resources

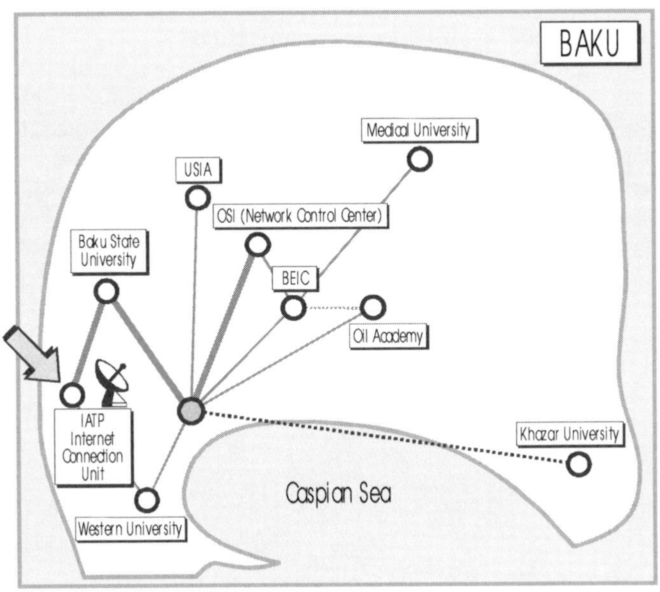

- *The Web Hosting* system is a huge storage space for content resources. Every content provider will get a limited space on the Web/Cache server.
- *The Central page site navigator* and search engine of national resources will facilitate the search for Azerbaijan related information;
- *The Public Mail* Server provides free e-mail access for all centers and individual users, running mail and news services;
- *The Remote access* system for dial-up users, which is granted with the Internet starter kit.

The Internet Connection Unit: Hosted by IATP, the unit is located at one of the highest places in the city, where the receive-only satellite dish was set up. Present connection to the local ISP is established with a leased cable line at a speed of 128 kbps. A receive-only satellite connection will be established via Netsat with an initial speed of 256 kbps and combined with the Internet connection to the local ISP.

Internet connectivity

The donor partners who pay for part of the connection allow other organizations to be connected, especially universities, schools, cultural organizations and local NGOs, to the network. As the network grows, it is possible to build a partnership of local organizations and gradually increase the bandwidth and their financial contribution to the connection.

Baku Information Network: We plan to use xDSL modems through existing leased cable lines at speeds no less than 128/768 kbps to connect the central communication unit with the Internet Connection Unit and Network Control Centre. CDMA radio-links will be used to connect remote centres in the poor-communication zones with the central communication unit at speeds of 128 kbps.

NETSAT Access. Plus receive only internet connection.
Dish — 3.7 m. Speed up to 768 kbps

Regional Centers: Regional information centres in the cities of Gence, Mingechevir and Lenkoran will be connected to the central communication unit via the dedicated phone lines leased from the local long-distance telecom company. The connection speed is up to 33.6 kbps. The Internet connection for the regional centers will be provided through DirectPC satellite up to 400 kbps.

The IATP Internet Connectivity Unit

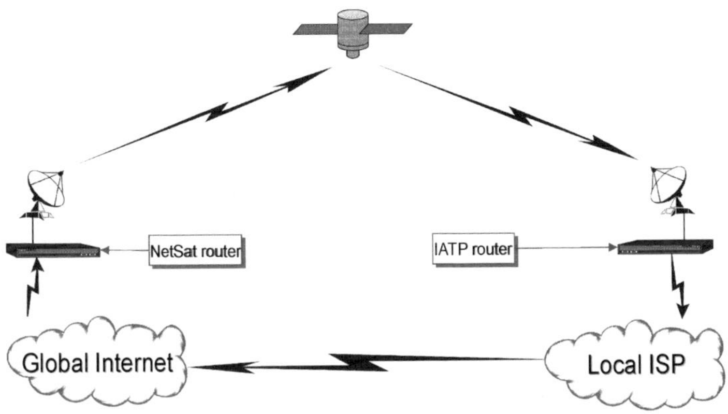

Asymmetric Satellite Connection Scheme

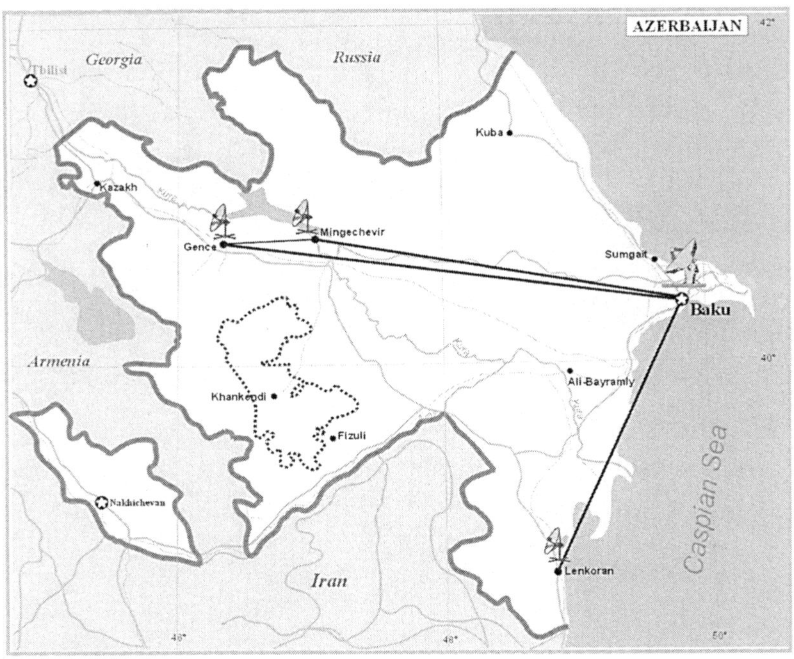

Future development

It is the intention of the partners during the year 2000 to oversee the establishment of a local non-governmental non-profit organization in Baku that will continue the development and maintenance of a non-commercial network in Azerbaijan. The structure should become a primary agent for providing free or low-cost Internet connectivity and online resources in Azerbaijan. We propose some form of Academic Internet Association, with each member paying a set fee, based on the capacity of their connectivity.

Divided among all organizations connected in this Project, the connectivity cost per organization *is far lower* than if donors or the organizations themselves had separate connections.

Private donors have been scared off by their experience with the Academy of Sciences, but we expect that, after the Network is developed, it should become more advantageous for them to provide further support, since success breeds success.

Internet resource training centers

General description

The Center possesses 10 terminals, which provide free access to e-mail and the Internet. After 1.5 years of operation, the center has over 3200 registered users, each of whom has a private e-mail account.

IATP training center

OSI training center

The Training Center provides 4 groups of training on Internet technologies and Web content creation:

- *Basic* Level for Internet *users*;
- *Intermediate* Level *non-professional Web-content creators;*

- *Advanced* Level for *Web designers;*
- *Developers* for *Web professionals.*

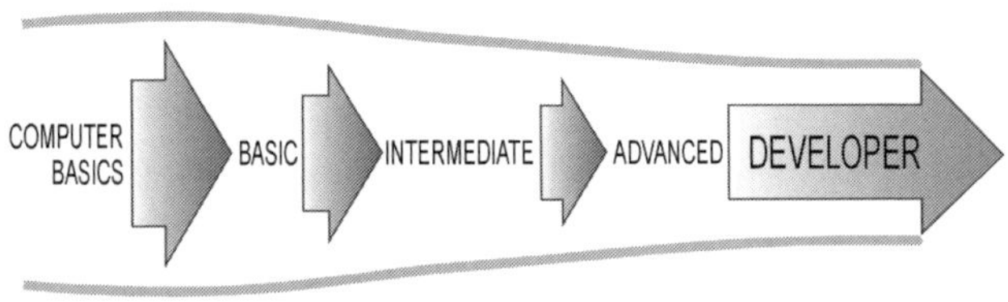

Internet Resources Training Centres

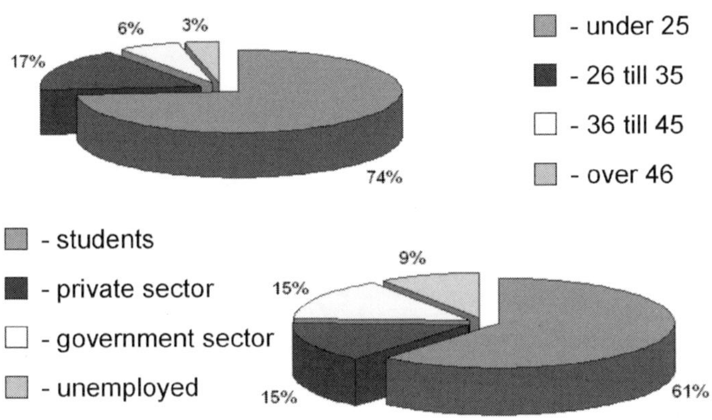

User Statistics

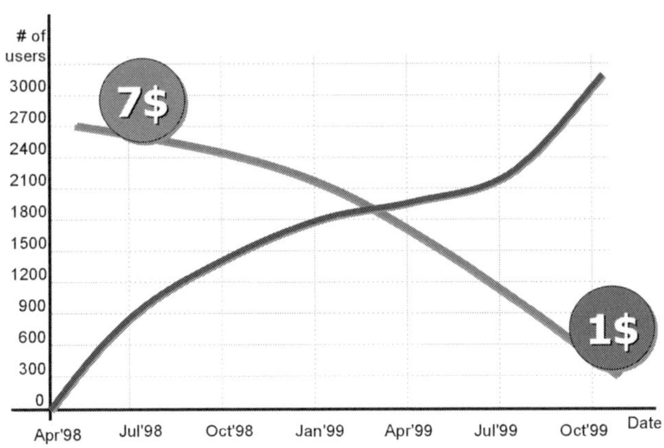

Internet Market Expansion

Trainings are organized in a funnel manner, i.e. higher levels assume smaller numbers of participants.

A statistical survey on attendance to the Center showed the following figures: students (61%), private sector (15%), government sector (15%), and unemployed (9%). Although the activity of the Center is rather attractive to all levels of population, students and the young generation remain the most active. This can be clearly seen from another survey on the basis of age: under 25 (74%), 26–35 (17%), 36–45 (6%), 46 and above (3%).

By providing comprehensive trainings courses, Internet Training Centers are expanding the market of potential users and drawing different sectors of the society into Internet technologies. The growing Internet community initiates demand for an infrastructure development in the country and expands connectivity, making it cheaper.

Jacek Gajewski

CEENet and the South-Eastern European Region

This talk is just a short introduction to the following three talks by the representatives of MARNet, CARNet and ARNES.

Central and Eastern European Networking Association (CEENet) is an association of academic, research and educational national network organizations from 23 countries; it streches from Estonia to Albania and from Austria to Mongolia. In particular, the following countries from the Caucasus Region, Armenia, Azerbaijan, Georgia and Russia, are members of the CEENet.

CEENet's main mission is the international coordination of the establishment and operation of NRENs. This mission is accomplished by the following CEENet activities:

- Information exchange during meetings and conferences;
- Training, workshops, conferences;
- Publications;
- Project definition, planning, management and realization.

CEENet has already worked intensively with the Caucasus Region, mainly by training young networkers. Since 1995 students from Caucasus have been coming to our workshops and they are more numerous than the average from other countries. Due to the recent changes of CZ, HU, andPL being accepted as NATO members, citizens of those countries can be NATO project coordinators. In particular, Jan Gruntorad (CEENet Chairman) is a co-director of a NATO funded infrastructure project in Azerbaijan.

CEENet can also help Caucasus in a project partner search - CEENet is officially registered under the Austrian law and is an eligible partner in EC projects within FP5 program. There was a lot of information exchange between the CEENet community and the Caucasian networkers during the previous network policy workshop in Tartu, network management workshop in Yaroslavl and the CEENet General Assembly meetings.

We would like to help the Caucasus Region as much as we can. That is why we have suggested to this workshop Organizing Committee that a special session be devoted to the Balkan Region experiences. We believe that South-Eastern Europe, and the Balkan Region in particular, is in some aspects similar to the Caucasus and the experience of Macedonian, Croatian and Slovenian NRENs might be useful.

The following similarities between Balkan and Caucasus Regions might be important for the creation and operation of NRENs:

- Countries from both regions have gained their independence through the dissolution of a multinational union or federation;
- There are still tensions between those countries;
- Geographically they are mountainous countries;
- The role of network education is well understood.

Networking Developments in the Caucasus Region
J.-P. Nadreau and O.B. Popov (Eds.)
IOS Press, 2001

Oliver B. Popov

MARNet: The Making of the National Research and Education Network or the Road to Utopia

Introduction

In the last ten years, it became acutely clear that both networking and global connectivity (or the access is to the Internet) are of a strategic importance for all country in the world. This is even truer for those that want to transform themselves from the "have-nots" status to the "haves" (the phenomenon of the convergence to development).

Very often the implementation of this crucial task is undertaken by a single national organization which ordinarily emanates from a research and education environment (Geof Huston calls it "the grand experiment"). The concept is consistent and complementary to the one that made Internet-style networking possible — sharing the best. It proves to be a decisive factor in the realization of a sometimes difficult yet highly noble endeavor — the right to communicate globally for everyone.

Similar beginnings

As in many other countries, initial networking expertise was developed and disseminated by the departments of computer science. It was also fostered by a need for fast exchange of data and computing power by the scientists working in chemistry and physics. This pattern of behavior reflected in the buildup of university networks throughout the former Yugoslavia. In addition to the basic internetworking within each university, the campus wide networks were to supply the communication infrastructure for the integral information system to connect the relevant databases, which comprised data about researchers, scholars, research and application projects, technological and industrial innovations.

The underlying computing technology within each University was a blend of IBM and Digital machines, while the intersystem connections were resolved through the DNA and SNA protocols. All of the universities were connected via X.25 a protocol well implemented through the unified PTT system termed as YUPAC. Yugoslavia was one of the first countries in the CEE region accepted in to EARN, hence becoming a member of the world wide area network BITNET.

The link to Macedonia went down in May 1992, with the exclusion of Serbia, as part of the whole system of international sanctions, naturally, a result of the war ignited in some regions of what used to be Yugoslavia.

The connection was restored by the end of the same summer, via a single X.25 link to the University of Ljubljana. This proved to be a rather costly affair, with only limited services (an e-mail and an option for a Telnet to a remote machine for a few that had accounts).

The Internet presence

Slowly, and mostly through scholars with a "foreign" experience, Internet triggered some interest in Macedonia. Members of the academia prompted the Ministry of Science to establish a Steering Committee that was supposed to:

- Investigate the most efficient way to connect to Internet, and
- Study and recommend the legal framework and appropriate institutionalization of an entity responsible for research and educational networking.

It was a modest step towards Internet connectivity. However, the Committee assisted in the mobilization of certain factors. Looking back, after an objective evaluation of its role, the Committee had a profound, yet limited impact on the whole affair due to restricted options, dubious expertise of certain members, and selective and channeled political support.

One of the issues addressed by the Committee was membership in the European networking associations (RARE and EARN). EARN accepted MARNet under the title of "University of Skopje", while RARE rejected its application again due to the EU political restrictions againt Macedonia. With the merger of the two entities in TERENA, the membership was from EARN translated into the new association.

The major result was the Pilot project for MARNet, submitted to several international organizations for revision and evaluation, and subsequently unanimously approved. In the second half of 1993, MARNet was invited to joint CEENet, a networking association in the process of establishment.

ACONet, the academic and research network of Austria, already had well organized programme of support and an assistance for most countries in the CEE region, which was also extended to Macedonia. ACONet financed the cost of a terrestrial half-link form Skopje to Vienna, which became later through the ENS a point of connection to Internet. The funding provided by the Government of Austria through ACONet was the catalytic factor for other national institutions and organizations to join the whole endeavor. Hence, the Ministry of Science and OSI in Macedonia approved the additional funding.

The University of Skopje, which had already built an internal network structure, while rudimentarily based on DECNet and VMS OS environment, slow lines, and almost nothing built on the application level, became, the physical and logical site of MARNet. The almost non-existent communication between various departments and colleges at the University experienced a renaissance with the possibilities offered by the full Internet. The MARNet international connection became operational in June 1995, via VSAT link to Vienna at 64 kbps.

Current situation: alive more or less

For the last six years, since its de facto inception, MARNet has been rather active in CEENet, ISOC, and yet rather marginal in TERENA. This is due to the structure of the organization and the voting rights (which are understandably quite unfavorable for small and not so advanced countries). MARNet has played a pioneering role in the educational community, primarily at the university level. Yet, the existence of the network made the decisive threshold for overcoming the basics of Internet and network literacy in the country, whose network culture and ethics is still in its inception and undesirably low. However, there are signs of an improvement in a positive direction, which also mirrors the change of attitudes.

The principal objectives of MARNet, to provide connectivity both on a national and international scene for the academic and research community, to promote the national identity of the country, to introduce and explore new CIT, are still the same. But very little has been done on the organization level. Namely, it is a separate unit within the University of Skopje network, a semi-legal entity with no clearly defined status. The number of employed people has not increased from the start, and finally there has not been any improvement in the capacity of the international link. There has been some outsourcing of the required services, but it has been carried out in an incidental manner.

In the last two years, the second university in Macedonia, the University Kliment Ohridski in Bitola, developed and implemented a state-of-the-art wide-area campus network, based on the distributed computing environment paradigm. Consequently, another significant part of the academic community gained an access to the Internet. Also, significant equipment was donated to MARNet via the PHARE programme, although the distribution was a little bit questionable and does not merit the University of Skopje authorities.

By the end of 1998, MARNet had two international links, the old satellite link partially funded by ACONet with acapacity of 64 kbps, and a new one through the Macedonian Telecom which alternated its capacity between 256 and 512 kbps. However, the satellite link through Austrian Telecom became unreasonably expensive, despite the excellent service including maintenance. Moreover, crisis in Kosovo in spring 1999 had a severe negative effect on the overall budget of the country and the one planned for MARNet. It made it impossible to continue this link, which was finally disconnected in October 1, 1999. At present, only one link of 512 kbps through the Macedonian telecom is operational.

It appears that the new telecommunication law, instead of deregulating like in the rest of the free market world, even more then before, regulates and encourages the monopoly of the NPO. All of the commercial ISPs, whose number surged to six, had to abandon their physical links and go through the Telecom. However, there is an indication that the internal infrastructure in the country is improving, and moreover the Telecom has been rather supportive of MARNet, unlike in the past, by providing relatively favorable prices for their services.

A bright star that emerged from the MARNet activities, and originally its major financial contributor, was the network developed by the Open Society Institute of Macedonia, a part of the Soros foundation (which played a pioneering role in the proliferation of the Internet in the CEE region and beyon). For the first two years, most of the cost for the link, memberships in international organizations, and equipment was covered by

OSIM. Later they established their own link via the Telecom, however there is a peering agreement between the two entities, MARNet and OSIMNet. The OSI

- Gave free service to the other NGOs in Macedonia for a certain period of time;
- Introduced three more points of presence on other locations throughout the country, and Cyber Corners for young people;
- Connected to medical institutions and individuals from the academic and arts communities;
- Penetrated secondary and primary education by providing connectivity, equipment and education for the teachers;
- Sponsoring numerous projects in arts, history and folklore;
- Provided additional education for four hundred teachers and librarians, as well as Web hosting for museums;
- Financed Internet education for Roma, Kosovo, and women related NGOs;
- Established BB services for the refugee camps in Macedonia and logistic support for the Kosovo-based OSI organization.

The OSIM network, in the spirit of self-reliance and sustainability, is slowly moving from free to fee for service. However, the quality of the work and the reasonable fees still make it the most appealing net provider, in addition to MARNet, for target communities such as education, culture, health and NGOs.

The road ahead of MARNet

One of the major obstacles in aperiod of political and economic change, as an undesirable side effect, such as dubious ethical standards (lack of willingness to share, divide, co-operate) still plagues the real development of the Net. This type of culture is reflected in the fragmentation that exists within the University, where the integrative forces have a very obscure and restrained power over any other individual constituent (a sort of slight perversion of individual rights). The existence of instability and political unrest in the neighboring regions does not help either.

It is not all about money. Very often, money can be found, but the problem is how well it is going to be spent, that is distributed and allocated. While the numbers are different, financial problems are similar to the ones in developed countries, and these types of problems are inevitable and part of the milieu.

MARNet has overcome (more or less) the inadequate behavior of some "commercial" providers they used to be commercial with the respect to charging, but certainly not with respect of quality to service and consistency, at least in the beginning. Today, the seeds of competition have driven the prices down and raised the quality of service.

One problem remains, that is, the divisions among the highly decentralized University, which may be resolved with the new law for Higher Education.

MARNet should also join, like other European networks (ARNes and CARNet are possibly the best models for us), the TEN-155 (Quantum) project or its follow-ups. One possible venue is to establish with Albania, Bulgaria, Greece, Romania and Turkey, a Balkan Academic and Research Networking Association (BARNet). The membership of

Greece in the EU and its most advanced networking situation in the region, might improve both the organizational and infrastructure situation in the Balkan countries.

Indeed, MARNet is unqualified proof that co-operation between the government, NGOs, and foreign assistance along with a visionary volunteer effort could work. MARNet has tried to overcome ignorance by not charging for connectivity, but there are still problems with people understanding the technology and its impact.

The recent flood of renewed interest by the Ministry of Science in finally resolving the status of MARNet and exploring models for its institutional self-sustainability is promising. The moves are encouraging, although not fully clear since there are some indications about the unwarranted position of the PNO, which might evolve to a single ISP, due to favorable treatment by the present telecommunication laws (hopefully it will be modified if the country wants to join the EU). Indeed, the most difficult problem for all the people involved in the MARNet endeavor has been knowledge, adherence to and enforcement of European standards and values in the local situation. It has been the major source of frustration.

MARNet, both as an idea and then as reality, since the very beginning, was placed on sound and rather prophetic grounds. This reality is still alive, especially if one considers that the whole operation was and still is run with very limited manpower. It is also a proof that unique individuals with the right support can make a difference. Hopefully, this will result in a permanent presence and sustainable existence.

Networking Developments in the Caucasus Region
J.-P. Nadreau and O.B. Popov (Eds.)
IOS Press, 2001

Vesna Vrga

The Croatian Academic & Research Network — CARNet

Croatia

- A small and young country;
- With 4.7 million inhabitants, and
- 54,000 km^2;
- More than 1000 islands;
- Became independant in 1991;
- $\sim$ 100,000 Internet users.

Croatian academic community

- $\sim$ 90,000 students;
- 7000 scientists;
- Four Universities (Zagreb, Split, Osijek, Rijeka);
- 90 Faculties and High Schools;

- 25 Institutes and Hospitals/Clinics;
- Ministry of Science and Technology.

Project CARNet

- Started as a project of the Ministry of Science and Technology in 1991;
- Main goal — to connect the Croatian national network to global information systems;
- 1991 — TLD.hr and connection to EBONE (Vienna);
- 1995 — Government founded CARNet as an institution.

CARNet as an institution — goals

Usual (for A&R network)

- to develop and to maintain a modern computer network for the academic and research community;

Less usual

- to provide intensive support to member institutions, as well as to end-users in the use of modern IT;
- to improve the quality of the national information space;
- to improve the quality of network (IT) use at the academic and national level;

Unusual

- to accomplish as many experimental implementations of IT as possible in different fields of human activity;
- to be a testbed for large national systems;
- to promote and initiate the use of IT in Croatia.

CARNet as an institution

- Centrally financed: state budget ($3,14 M in 1999);
- 35 employees (average age: 28);
- Three offices in Zagreb + four offices in regional centers;
- Flat organizational structure and project oriented;
- Offers services, apart from the academic community, to all citizens of Croatia, and
- All services free of charge.

Network infrastructure

- Well developed to meet today's and future users' needs;
- 1995 — first ATM connection (155 Mbps);

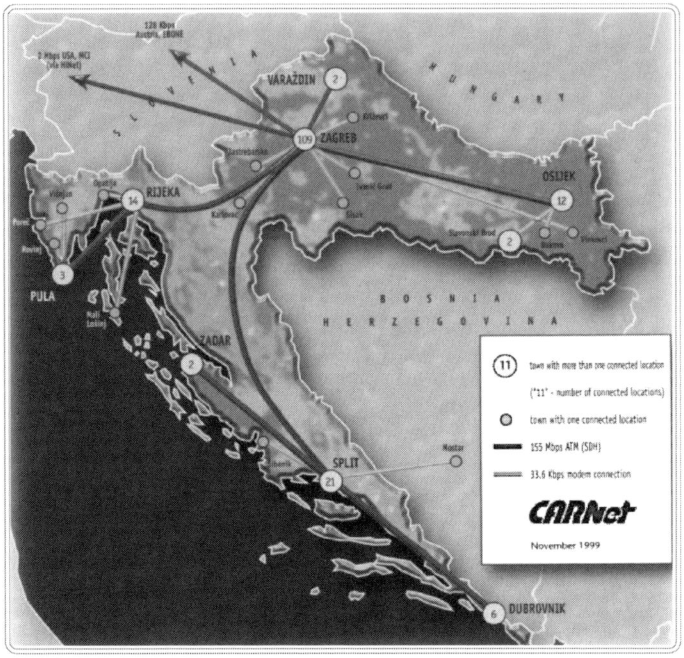

- 1997 — ATM Backbone Core at 622 Mbps (7 locations within Zagreb);
- Relaying on the Telecom infrastructure;
- Four universities;
- Eight cities in the Backbone — ATM 155 Mbps;
- 23 Cities;
- > 150 member institutions;
- > 180 locations;
- 2 international links:
  - 4 Mbps to USA (MCI),
  - 2 × 64 kbps to EBONE (Vienna).

User oriented ...

- Member support;
- Help desk, courses, seminars and educational materials
  - for end-users (more than 1.200 courses, more than 20.000 participants),
  - for future teachers (since 1998, more than 30 teachers qualified),
  - for system administrators;
- Dial-ins and public hosts;
- Information services;

- Videoconferences and distance learning;
- CARNet weeks, users conference and event support;
- CERT & DNS.

... And focused

- Carried out in cooperation with the academic community;
- Infrastructure projects;
- Research and development projects;
- Reference centers (VRML, Mathematica, Online Data Base, ...);
- Information services;
- Pilot projects (astronomy, medicine, culture, ...);
- Examples of application of IT in various fields.

Changes ... and ... future

- CARNet is changing its role from ISP to IT project consulting agency (RKMTDU, NISKA, LANs, CROSNet, ...);
- Maintenance and services outsourced;
- More emphasis on educational systems and consulting.

Marko Bonac

The Academic and Research Network of Slovenia

1. Introduction

Simple access to a variety of information and cooperation with related organizations are of crucial importance for the success of research institutes, universities and other organizations involved in development, education and culture. This can be achieved today with computer networks. Since there is a big difference between the general purpose Commodity Internet on one hand, and the High Performance Network that satisfies the needs of researchers and professors on the other hand, you can find so called Research and Educational Networks in all countries. These networks provide "production" level services to educational, research and cultural institutions and serve as the test bed environment for the demonstration and validation of advanced networking applications in the field of research and education.

2. Legal structure

The Academic and Research Network of Slovenia (ARNES) was established as an independent public institution in 1992. The sphere of activities, organization and funding of ARNES is regulated by the Foundation Act of ARNES as a public institution. The Act defines ARNES as a research and educational infrastructure for the development, organization and management of the academic and research computer network in the Republic of Slovenia.

ARNES is directed by a Board of Directors. The Board consists of six members, three of which are appointed by the Ministry of Science and Technology, two by the Ministry of Education and Sport and one by ARNES employees. The Board of Directors adopts an Annual Operating Plan and a Final Account in accordance with the Ministry of Science and Technology and the Ministry of Education and Sport.

The work and operations are managed by the Managing Director, who is appointed by the Board of Directors with the consent of the founder.

ARNES has a Technical Board which deals with user-oriented, technical programme and development activities.

Funds necessary for the execution of ARNES operations are obtained from the budget of the Republic of Slovenia on the basis of the Annual Operating Plan. ARNES can also

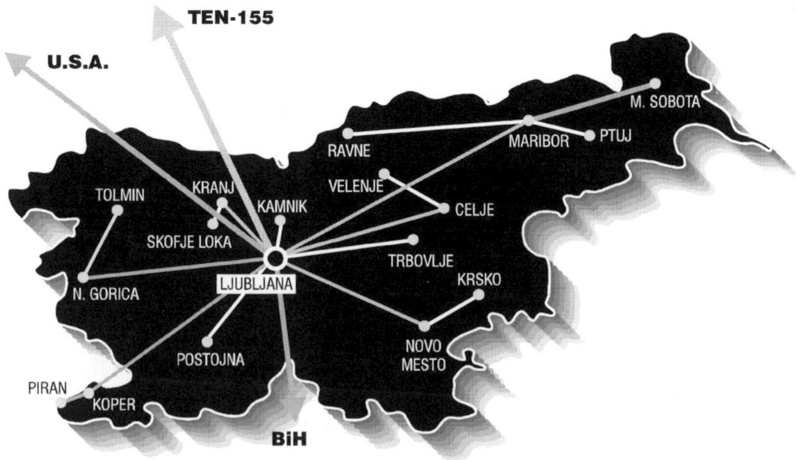

get funds from other sources. ARNES is required to use surpluses to carry out and develop its activities.

In legal matters ARNES acts independently.

3. Connectivity

3.1. Connectivity inside Slovenia

The basis for connectivity inside Slovenia is the backbone which is constructed by main routers (nodes of concentration) connected by leased lines. The ARNES network is composed of the ARNES backbone lines to it, and routers at customer premises. The whole network is managed by ARNES.

Nodes of concentration (NOC) are located in major towns in Slovenia: in Ljubljana, Maribor, Kranj, Celje, Koper, Nova Gorica, Novo Mesto, Murska Sobota, Ravne na Koroskem, Krsko, Postojna, Trbovlje, Ptuj, Portoroz, Kamnik, Tolmin, Skofja Loka and Velenje. These nodes of concentration are connected with leased lines. Small towns are connected with 256 kbps lines, larger with 512 kbps or 2 Mbps, Kranj with 10 Mbps and Maribor with a 155 Mbps line.

At present, 227 research institutions, educational institutions and libraries are connected to NOCs via leased lines. Most of these lines have a capacity of between 64 kbps and 2 Mbps. Some bigger institutions are connected by Ethernet (10 Mbps).

ARNES has 1520 dial-up ports in most of the towns for terminal or PPP-based remote access for those organizations from the research and educational area which cannot afford a leased line. Connection via public X.25 network or ISDN is also possible.

We estimate that there are altogether 40,000 individuals from 1200 institutions who are connected to ARNES network and use ARNES services in Slovenia.

3.2. International connectivity

ARNES has a 10 Mbps line to the Pan-European research network TEN-155 and 2 Mbps to the USA. An upgrade of the TEN-155 connection to 45 Mbps is planned for November 1999.

3.3. Ways to connect to the ARNES network

There are several ways in which an individual organization can connect to the ARNES network. The choice depends on the requirements of the organization, its local infrastructure and the funds available. The organization can have a permanent connection (via leased line or frame relay) or a dial-up connection (via analogue or ISDN telephone network). It is expected that more options will become available with the liberalisation of telecommunication (wireless connections, private optical fibres, xDSL, TV cables etc).

Individuals from research and educational institutions can get a personal dial-up account directly from ARNES. They can use analogue or ISDN telephone network.

4. Users

ARNES currently provides network services to universities, secondary and primary schools, government and private research institutions, libraries and cultural institutions.

Type of organization	Number of permanent connections	Percentage in dial-up accounts	Percentage in traffic
Institutes	39	7%	16%
Development in industry	19	4%	5%
Universities	7	5%	41%
Secondary schools	40	41%	17%
Primary schools	18	30%	9%
Other education	13	4%	4%
Libraries, museums	77	3%	5%
Other	14	6%	3%
Total	227	100%	100%

5. Funding

The main cost for ARNES comes from international connectivity, which accounts for about 60% of the total cost. The next largest sum or about 20%, is used for national connectivity. Purchasing of equipment takes an additional 11%. All other costs (salaries, equipment maintenance , insurance, etc.) amount to 9%.

ARNES is primarily funded by the Ministry of Science and Technology, and so is free for most users. The exceptions are development and research departments in industry — they must to cover the cost of the services (which is calculated on the basis of their incoming traffic).

6. Network services

One of the purpose of a computer network is provide its services. ARNES carries out all the necessary activities to facilitate the use of all Internet services.

ARNES also maintains its own WWW, FTP, WAIS, X.500, News, MBONE, IRC and Gopher servers. There is a possibility that those users who can not maintain their own WWW server be allowed to do so on one of the ARNES servers.

From the very start of networking in Slovenia, ARNES has supported some basic activities which are useful for other Internet service providers in Slovenia as well. ARNES manages the top level domain (.si) for Slovenia and runs the top level Domain Name Server. In addition ARNES runs SIX (Slovenian Internet Exchange) where all those Internet providers with their own international connectivity can peer between themselves.

7. User support

Computer networks are a new technology and organizations very often ask for technical support when they requesting the connection of their local networks to ARNES. Similarly, the services of the global computer network are new for most users. ARNES has to help here as well.

The activities which need to be carried out are the following:

- Consultancy to organizations on how to connect their local networks to ARNES.
- Consultancy to organizations on setting up their DNS, E-mail, FTP, WWW, News and X.500 systems.
- Organization of conferences for managers of local systems.
- Help desk for managers of local systems.
- Computer Emergency Response Team support (SI-CERT).
- Consultancy to organizations regarding use of services.
- Cooperation in seminars organized by connected organizations for their users.
- Organization of conferences for future/prospective users.
- Presentation at exhibitions.

8. International cooperation

ARNES is a member of TERENA, a shareholder in DANTE, a member of CEENet and a member of RIPE.

TERENA (The trans-European Research and Educational Networking Association) was established in 1994 through the merger of the former RARE and EARN associations. The mission of TERENA is the promotion and development of a high-quality international information and telecommunication infrastructure to support European research and education. TERENA is an association and its members are national research and education organizations from Europe. TERENA brings the interests of its members to the attention of governments, funding bodies, regulators and industry. TERENA also organizes conferences and workshops and provides certain services when they are not yet mature enough to survive on their own. The central activity of TERENA is its technical programme which provides an impetus for the development, testing and deployment of new technologies in the research networking environment.

DANTE (Delivery of Advanced Network Technology to Europe Ltd.) is a non-profit, limited liability company with European National Research Networks as shareholders. DANTE organizes, manages and provides advanced pan-European data network services to the European research community. The main service is pan-European connectivity provided by the TEN-155 network. Its core consists of 155 Mbps circuits and is supplemented by 34 and 45 Mbps circuits to smaller countries. TEN-155 is based on SDH technology with an ATM overlay. Much of the capacity is devoted to a continuation of a "best efforts" IP service, but the ATM overlay also provides for a Managed Bandwidth Service to be offered to specific groups of users and for the temporary setup of virtual paths with guaranteed bandwidth between national research networks and their user organizations. TEN-155 network is established as the EC cofunded project called QUANTUM.

CEENet (Central and Eastern European Networking Association) is the association of National Research and Educational Networks from Central and Eastern Europe. It has twenty-one members, from Austria to Mongolia. Its main activities are the organization of a series of educational workshops. The generous awards and grants from NATO Science Committee Programme and Soros Foundation provided a sound financial basis for the education of more than five hundred individuals from the CEE countries. Apart from technology workshops there were also workshops on network management and policies. The foundation of CEENet has proved to be a very important factor for the inception and improvement of many national networks in the member countries.

RIPE (Réseaux IP Européens) is a collaborative organization open to all European Internet service providers. The objective of RIPE is to ensure the necessary administrative and technical coordination to allow the operation of a pan-European IP network. The RIPE Network Coordination Centre manages the databases on IP networks, DNS domains, IP routing policies and contact information. RIPE NCC also distributes IP numbers in Europe.

ARNES also participates in many European projects in the field of network technologies and network services.

9. Regulatory development and liberalisation evolution

Regulatory development and the degree of liberalisation in the telecommunication field plays a decisive role in the development of Internet. Telecom monopoly is the main obstacle hindering a faster growth of Internet in some Central and Eastern European countries.

9.1. Situation in European Union countries

In 1986 the European Commission published its famous Green Paper which started the first wave of liberalisation in Europe. Most countries opened the telecommunication sector to full competition on 1.1.1998. Regulators try to develop rules that can assure fair competition and a wide choice of quality offerings for users.

Competition regulations are covered by several EU directives. They can be divided into the following groups:

- **Liberalisation:** these directives determine fair competition in the telecommunications sector and are interpretations and extensions of normal competition law.
- **Harmonisation:** these directives try to ensure that the basic telecommunications regulations are the same across the European Union.
- **Independent national regulator:** the regulator has to be independent and have enough authority to decide and enough power to enforce the decisions on the licensing of new operators, numbering and frequency management and dispute resolutions in interconnection conflicts.
- **Licensing:** should be an open and transparent process. There can be no limit on numbers of licenses except in the case of limited resources or where there are special delivery obligations for organisations with significant market power. Most services should be able to be provided without the need of an individual license but on the basis of a general authorization.
- **Open network provision:** is a set of provisions aimed at ensuring fair competition in an environment where there is very uneven market power.
- **Universal service:** is defined as the provision at a reasonable cost of some basic telecommunication services for all.
- **Data protection:** defines how the operators must protect and not misuse personal data.
- **Type approval:** defines type approval procedures.

9.2. In Central and Eastern European countries

The majority of CEE countries have chosen to delay full competition until a date between 2001 and 2003. A period of preparation is sometimes necessary to adjust tariffs to the underlying costs. Such changes are not always easy to introduce overnight and have to be implemented gradually in order to be politically acceptable.

Most of the CEE countries go into regulatory reform because they want to become members of the European Union and have to meet the “acquis communautaire”. Sometimes the politicians in these countries are not happy and believe that this is the price to be paid in order for the access to Western Europe. In addition, foreign investors are sometimes willing to pay more for shares in incumbent operator with monopoly rights, because they know they will get this money back through higher telecommunication tariffs and make higher profits.

CEE countries must decide how to arrange telecommunications to create maximum competitive advantage for their economy. In many cases the answer is that it is not wise to artificially prolong the monopoly period. They could get a better price for their incumbent operator with monopoly rights, but this price will be paid by all users in the country. In addition to the monopoly situation the country cannot benefit from all the new technologies that are becoming available and which may dramatically reduce the cost of telecommunication networks.

9.3 Slovenia

The new Law on Telecommunications of May 1997 represented a step toward a competitive Slovenian telecommunications market. However, there are many areas where the regulation is rather restrictive:

- The regulatory authority does not meet all the criteria for an independent regulator;
- The competition authority takes an active interest in telecommunications but is dependent on the slow court system for sanctions;
- Licensing procedures require individual licences where a general authorisation would be enough;
- In interconnection issues, no distinction is made between operators with significant market power and other operators;
- There are no funding mechanisms from new operators for the universal service. Instead budget funds are mentioned, which is perceived as a subsidy that can distort competition;
- While domestic PSTN tariffs are low, international PSTN tariffs are very high;
- International leased lines are very expensive. There is no obligation concerning the correlation between costs and prices of leased lines. The Minister responsible for communication can even decide that the incumbent operator is not obliged to lease a line, if this can have an impact on his commercial status;
- The incumbent operator has a monopoly in providing telecommunication networks for data communication and Internet.

10. Conclusion

The main task of ARNES is the development, operation and management of the communication and information network for education and research. The objective is to promote and enhance national and international cooperation in the field of teaching and research with the most up-to-date means of communication available. ARNES closely follows the progress in protocols and services in this area, actively promotes international standards, and avoids being limited to any single computer and communication supplier. Priority is given to services and user support.

Networking Developments in the Caucasus Region
J.-P. Nadreau and O.B. Popov (Eds.)
IOS Press, 2001

Predrag Pale

Information Technology, Academic Community & Future: ARNet Model

The main idea behind this presentation is to broaden views, to show what you could do yourselves, what your roles might be and make you aware that everything you decide to do is just a part of the big picture you might be thinking about or turning away from.

After that, all kinds of problems related to building and running an academic network will be presented, so that you do not think that the most important thing is to raise money and buy something.

Furthermore, it is easier to go through the entire process if you can get help from someone who has already gone through it.

In the end, possible solutions will be offered.

The role of academic community and IT in it

- **Academic Community**
  - autarchic?
  - leader in development (of society)?
- **Information Technology (in AC)**
  - a support system?
  - an infrastructure?
  - a tool?

We can start talking about problems and solutions in the implementation of IT in an academic community only after we have set our goals.

Setting goals starts with thinking about the role of the academic community in a society. Should it be autonomous, that is self-sufficient, and do scientific work for its own purposes, close up and act as if the rest of the society did not even exist? Or should it go to the other extreme and take over the leading role in guiding the society into a brighter future, set an example and help others find new ways of solving old and new problems?

When we figure this out, we should see if IT in the academic community is only one service or if it is a whole (rich, diverse, intensive, active) infrastructure or a mean of achieving (higher) goals.

Apparently, "our model" is an answer to both problems: an academic community which will rapidly lead everything else forward, with IT as a mean to do o. This does not mean that the tool is not complex and constitutes a world of its own, but it is important to understand its role.

The role and activities of an ARNet

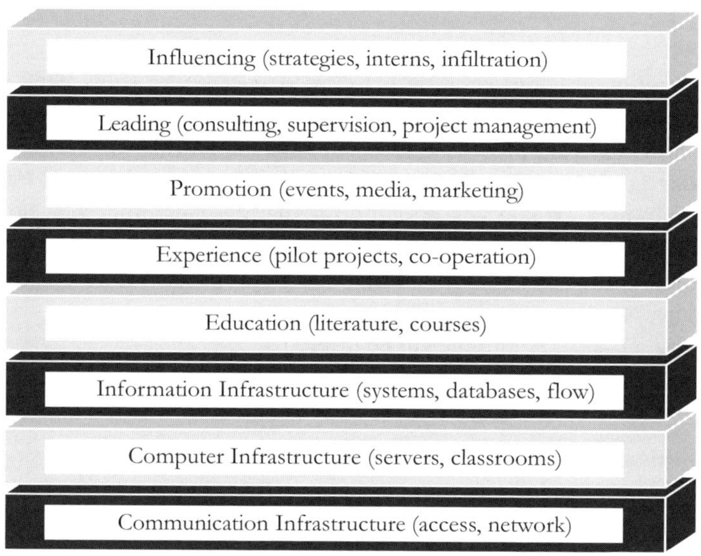

Here we wish to explain what an "Academic and Research Network" encompasses, because there are as many different opinions as they are national academic networks in the world.

We should discuss all types of activities.

The bottom line is the communication infrastructure. Most ARNets still believe that this is their main purpose. Some achieve this through one server with modem entries, and some build a large infrastructure which in rare cases reaches institutions.

Computer infrastructure means that we ensure that our institutions have (at least) one server and that it supports basic services. We also make sure that each institution has a computer classroom and other places with terminal equipment to access the network.

Information infrastructure also means providing databases, library systems, andall central servers (ftp, irc, directory services, DNS etc.) for the academic community.

Education can be left to schools, but they are (including universities) too inert. Commercial schools, however, can survive only by abiding by the laws of the market and this is nothing new. Therefore, some academic networks (like CARNet) educate their users, system administrators and trainers (TtT), and their own employees. The education consist not only of courses but also of literature and even promotion.

The experience of APPLICATION is extremely important for achieving high goals.

Promotion is important as a part of education and encouraging users to use everything at their disposal as well as help IT and AC become a part of our everyday lives. This is also a way of obtaining the necessary resources.

Leadership is important for major changes in a society. It requires knowledge other than the knowledge of the technology, for example project management. By consulting and supervising national projects you can guide development.

Finally, and that is the highest level, you can influence the direction and intensity of development of a society through preparing national strategies, employing interns from other parts of government administration and economy and infiltrating by sending your employees (temporary or long-term) to key positions in the government administration.

Problems

- Financial;
- Technical;
- People;
- Users;
- Political;
- Development.

Regardless of the field or fields an ARNet decides to operate in, it can encounter a whole range of different problems. Here I have listed only a few and the illustrations will follow.

I deliberately excluded organizational problems because they are embedded in all the problems listed.

Financial problems

- Planning;
- Stability of sources;
- Multitude of sources;
- Continuity;
- Growth;
- Tracking cost.

Financial problems start with planning. The amount of money and the dynamic is not always easily determined. Namely, if we do not manage to invest wisely, we shall probably have difficulties in obtaining the necessary resources in the future. If we spend it just so it does not get wasted, without a plan, sooner or later it will bite us in the neck. Of course, if we plan too little money, it will stop the development of projects.

The source of finance must be stable. Typically, money should come in even time periods. What can one do when the money starts coming irregularly or when the plan is not being executed?

The ideal situation would be to have multiple sources of finance. The danger however is that one source always expects that the other source covers expenses for a particular type of costs. Also, sources can often be incompatible (government vs. economy), that is, that the resources are used for a purpose that is crucial to the other source.

In each financial cycle (year) it is always a problem to obtain money from the same sources.

Especially if it is connected with growth, and when each year (apparently) the same job requires more and more money.

Also, there is always the question: "Would it not be cheaper to buy all these services on the market?" or at least "How much would it cost to double this part of our activities?" This requires that we, like commercial companies, keep track of our expenses through either cost centers or activity based costing. This is a complex system to set up and maintain which it is time consuming, money consuming and people consuming.

Technical problems

- Equipment/vendor choice;
- Compatibility;
- Maintenance;
- Obsoleteness.

Technical problems start with the decision about which equipment to buy, which vendor and distributor to use.

Which technology to use? Which technology will become obsolete and which will survive? Then there is the question of compatibility with what we already have, what do our partners have and what will be up-to-date tomorrow.

Maintenance is a separate problem. Vendors often require payment for maintenance, which is very money consuming, especially as the system grows.

A special problem is that the equipment becomes obsolete before it is really too old too use. A good example are modems and personal computers. Modems should be exchanged every 12 to 18 months and PCs every 24 to 36 months.

This is a huge expense (to exchange existing capacities) and an enormous psychological impediment (to "throw away" almost brand new equipment).

People problems

- Where to find them;
- How to keep them;
- How to grow them;
- Internal conflicts;
- When we become too big.

Nothing can be achieved without people.

Where can one find good experts, how to attract and keep them when the commercial companies always pay IT experts and good organizers better.

Besides, however capable people may be, they have to be constantly trained. If by no other means, then through access to specialized magazines and literature and new technologies to "play" with. But you cannot do without the "privilege" of education in the form of courses, in-service training, post-graduate studies etc.

Even if we hire good employees or set up good teams, conflicts among employees are inevitable. ARNets are usually led by people from the academic community or engineers who have underdeveloped social skills so the growth and maintenance of the community is a great challenge and strain for them and requires additional knowledge of these skills.

The greatest challenges emerge if we become very successful because this burdens us with a great deal of work and associates. The organization is growing fast and all of a sudden we become huge, inert, slow, like organizations we used to mock for those same things, that drove us crazy (Telecom, big companies, government administration).

User problems

- Why don't they use it;
- They keep asking the same questions;
- The more we teach them, the more they want to know;
- Now they think they can do our job better than we can.

We are aware of the fact that everything we do, we do for our users. Nevertheless they are like children: they know less than we do, often they do not comprehend important issues, persistion irrelevant facts and slow us down on our way to creating the "perfect" system.

They also usually use a limited amount of the services at their disposal.

It often seems that there is no end to answering basic questions and constant by repeating of basic rules and facts.

Although we constantly educate them, there are always new users who need to be educated and new courses that have to be prepared.

And when finally a critical mass of users learn to use the technology, they start being very "smart", that is, trying to manage our plans and activities.

Political problems

- Don't you see how important we are;
- Who controls us;
- Whom do we serve;
- Isn't it too important to be left to technicians/academics;
- Content !!!

Political problems are connected with the awareness of national decision makers, support, financing and competition.

Often we cannot understand why politicians cannot see how important the job that we, as an ARNet, are doing really is.

Eventually, ideas from the field of politics emerge (and they have little idea what IT really is and can) about what and how we should do our job.

Sometimes the question arisesas to who our users really are: only the academic community? Schools? Citizens? The state? Someone else?

When politicians finally start comprehending what it is really all about, they want to take complete control over the situation without fully understanding the issue. They think they will guide it better than technicians or people form the academic community.

Politics will sooner or later interfere with the content and information travelling through the system or in the system. The content sooner or later becomes an issue of great dispute.

Development problems

- What is our mission;
- "Rivals": to conquer, resist or retreat;
- Going commercial?
- Whose strategy is first?

An ARNet is anything but static. It is constantly changing and usually growing and spreading.

The first question is "What is our mission?"

In many countries there are several organizations that are doing an ARNet-like job. If there is no clear decision from a higher instance about the organization that will unite all these activities, rivalry and battle for money and status may occur on both national and international levels. Should we go into battle with our competitors, resist, merge or give up?

Each ARNet sooner or later has to deal with the following dilemma: whether to go commercial (at least partly)? It is important to understand that commercialization begins when the first service is charged, and it usually spreads like fire.

In the development of an ARNet there are various interests: those of the individuals financing it, users and the organization itself. Each of them has its own strategy. Which comes first when the strategy is being set for the next cycle? If this is not clearly determined we often find ourselves in a vicious circle in which time and money are wasted.

Mission impossible?

- "Your people"

  - start with young people, a lot of them,
  - send them out & train, train, train ...,
  - gradually insert some "old folks";

- The best among you should "go political";
- Do not go commercial;
- Project management is key to success;
- Support your users, it's all for them;
- Marketing is a question of survival;
- Think big;

– if you won't, nobody will.

When we look into all these problems, we get the impression that this is an impossible mission. Could a small group of technicians and people from the academic community change so many crucial issues in a country?

It sure can, because the greatest things in history started out as ideas of ambitious individuals.

The best thing is to divide problems into several small ones.

First you have to have "your own people". Start with young people, fresh from college and give them a chance and at the same time shape them according to the common mission. You should start with a large group from the very beginning. Give them experience and send them similar organizations where they can work (for a couple of months) and gain not only knowledge and experience but also a feeling of organization and corporate culture. Provide them with all necessary education.

A lot of young and inexperienced people can be like a nuclear fuel: when the critical mass amounts, it can explode. That is why you should gradually add older, experienced professionals attracted by youth and new technologies to the group and they will hinder blowouts of the young by their sense of balance and experience.

In order to ensure political support you should not wait for a wise and reasonable politician to appear, but select and educate the best among you to become politically active and create a space for your activities.

Do not go commercial, at least not in the first few years. Not even a little. The people from the academic community cannot do commercial things and conflicts with commercial sphere will only confuse your employees and users. Material benefits are usually insignificant compared to the final costs.

Big jobs are usually done as projects and project management is crucial here. Teach your employees how to do that.

Never forget to ask yourselves: "What do my users need?" and then give it to them. If you do not do that your work has no future and cannot survive.

Doing your job well is not enough. Your users, people who finance you, your future associates and partners should know that. Marketing is the key to your survival.

Although there are many immediate steps to take, never forget to keep an eye on the goal that you had set. And let that goal be big enough so that your effort is worth while.

Who can help you?

- Everyone!
- Send your people to work with them;

- Ask YOUR questions;
- Never allow gifts to destroy YOUR strategy.

Although this looks complicated it can be achieved. It is even easier if you get some help. Who can help you?

Everyone! Everyone that does a job similar to yours is a possible source of help.

Send your people to work with them. They will get help and your people will have access to information and knowledge and will gain their own practical experience.

Ask your partners, competitors, similar organizations to tell you what is bothering them, not what they have prepared for you as answers. Do not ask them how they do it, but rather how they would do it if they were you.

Beware of gifts! They often mean that you must do something according to the plans and interests of the donors and not according to yours. Never allow gifts to jeopardize your strategy only because it is better to have something rather than nothing at all.

What can CARNet do for you?

- Advice
  - technical, partnerships, organizational,
  - e-mail, phone;
- Education
  - interns, summer courses, in-house,
  - training materials & methodology;
- Consulting
  - projects, problems, policy;
- Project/policy design
  - infrastructure, service, strategy, vision,
  - organizational, national.

What, for example, can CARNet Do For You?

If you need advice or assistance of any kind (technological, organizational, partnership), CARNet will be more than glad to help you. Your people can get additional education or experience — through internships — working in CARNet for a few months or through short visits to CARNet or vice versa, employees of CARNet could visit your country. Furthermore, CARNet training materials and methodology can be customized to fit your needs.

CARNet grew in similar conditions and deals with similar problems as you do. Our eight-year experience can help you in making strategic decisions, building systems or determining policies.

Expenses related to such project can be divided, and possibly alternative sources of financing could be found, for example international organizations, non-governmental organizations or other national projects.

Apart from the pleasure of helping others, CARNet would in that case grow and accumulate knowledge and experience.

There is so much work to be done in this world that only by helping one another and exchanging experience and ideas, can we make our dreams come true and take the advantage of all these great possibilities.

Do you feel overwhelmed?

- It is nothing in comparison with thefeelings of your users and your politicians;
- The world is beautiful;
- The world of IT is beautiful;
- You have friends.

If they feel overwhelmed or tired by the list of problems and issues to deal with, possibilities and obstacles, let them worry, because their users and their politicians have an even bigger problem.

Let them think about how life and the world around them is beautiful and full of possibilities, how IT put us in a historically different positions, and how everything is easier if you have friends at your side.

Les Cottrell

Quality of Service

Overview

- **How do we measure Quality of Service**
  - Overview of methodology;
- **Problem areas**:
  - Generally,
  - How do Eastern Europe & Russia look;
- **How does it affect applications**
  - Bulk data transfer, interactive applications,
  - Loss, RTT, jitter, availability;
- **What can be done?**

Measurement mechanism

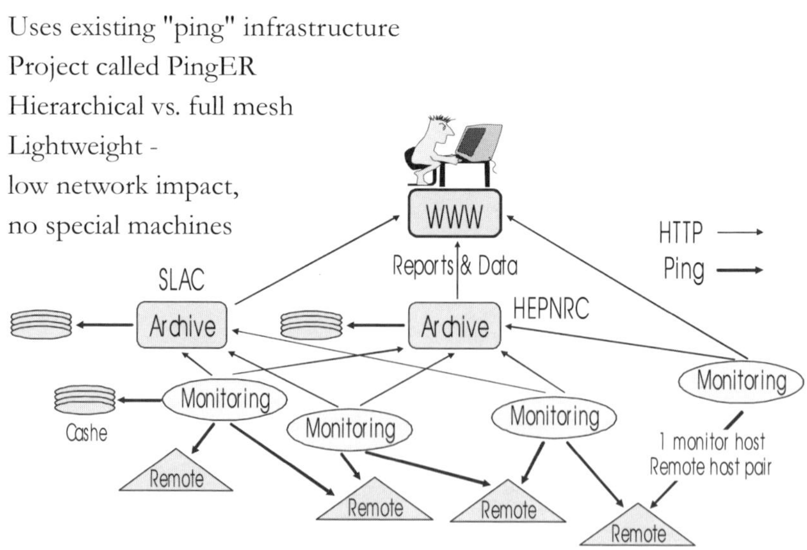

PingER worldwide deployment

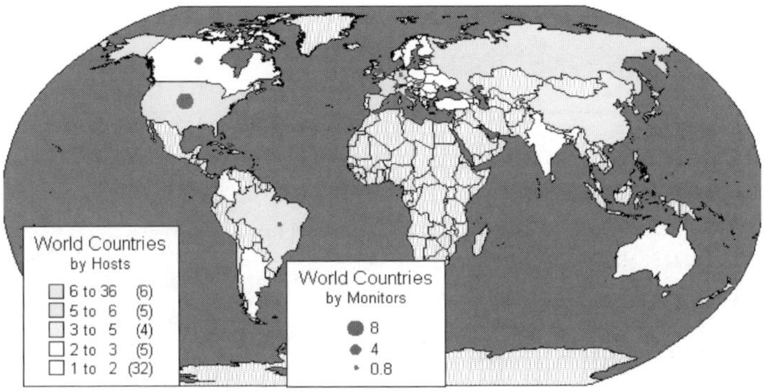

23 monitoring sites in 12 countries

511 remote hosts monitored in 54 countries on 6 continents

~ 2000 pairs

PingER deployment 2/2

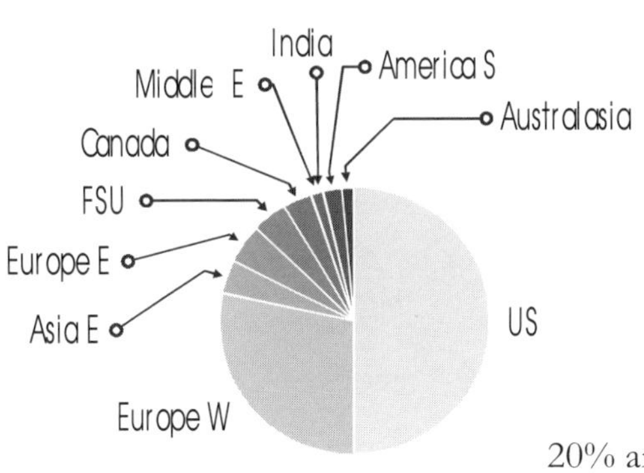

PingER remote sites
by area

In U.S.
57% are **.edu**
10% are **.gov**
15% are **.net**
10% are **.com**
20% are connected directly to **ESnet**
39% are on **Internet 2**

Results (Top Level View — Aug '99)

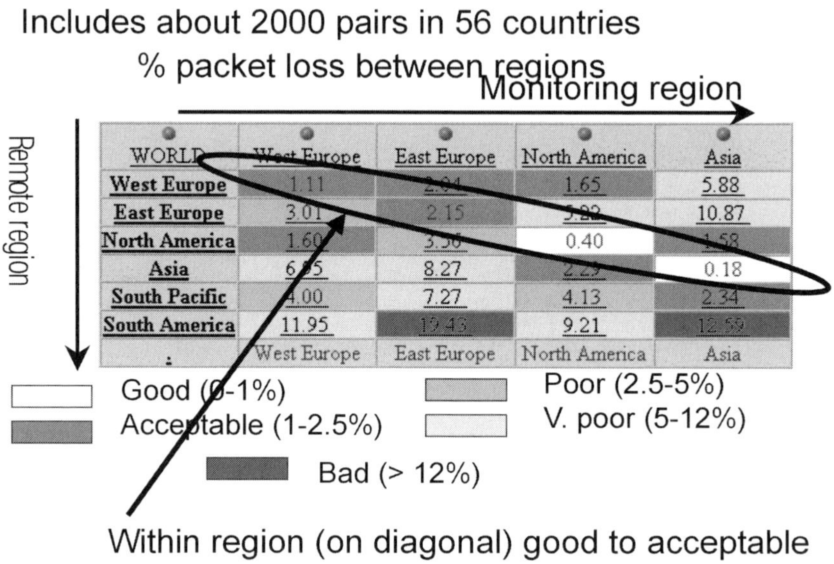

E — Estern Europe

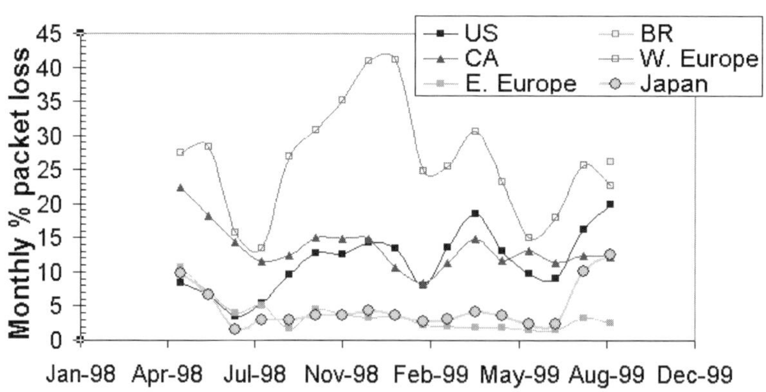

E — Europe Seen from the Rest of the World

Russia

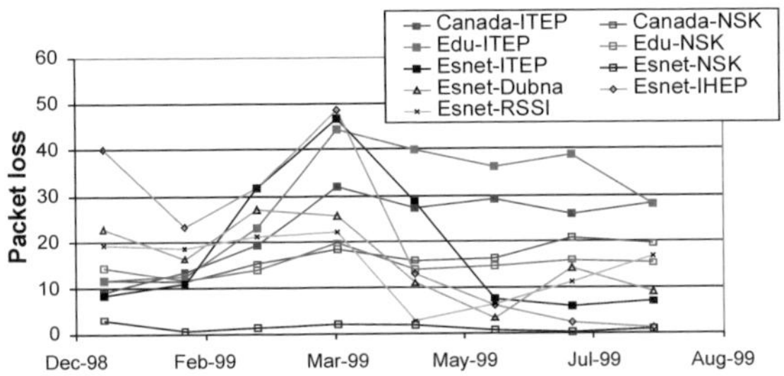

ESnet — NSk good, ESnet — ITEP & IHEP improved with new satellite
Canada & Edu bad all over
DESY, CERN improved to acceptable to ITEP, IHEP, NSK with new satellite,
Dubna still v. poor to bad, UK poor to ITEP & NSK, KEK good to NSk, v. poor to ITEP

Packet Loss from North-America to Russia (Jan-Aug '99)

European performance from U.S.

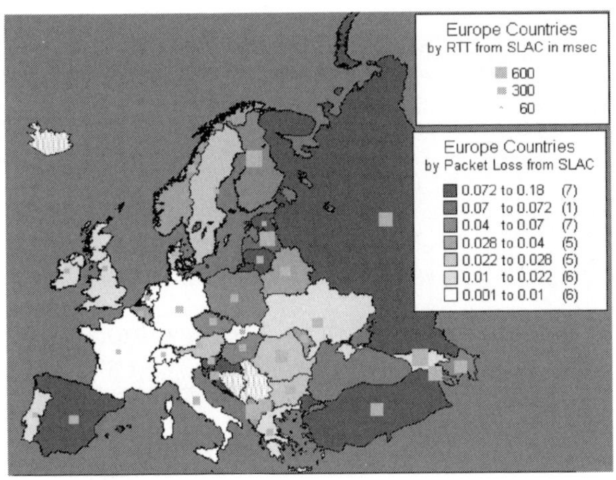

Europe Loss and RTT from SLAC

Problem areas

- **Germany** was bad with .ca & .edu yet good with ESnet. DESY improved in Aug with dedicated 3.5Mbps PVC to US/Canada R&E
- **Russia (W)** bad to .ca & .edu, good to ESnet, mixed to Europe, poor to .jp/ Dubna worse than others. ITEP/IHEP better since new satellite
- **E. Europe** generally poor to bad

- **China** poor to very poor with most
- **S. America** poor to very poor

Impact on applications

- E-mail
  - fairly insensitive to quality, may be delayed but keeps retrying for days and eventually gets through

- Web
  - usually has human but expectations are low, performance often more limited by server, can retry
- Bulk file transfer
  - unattended, if >10-12% loss connections can time out

- Interactive telnet, voice
  - very time & loss sensitive
  - E.g. telnet/ssh loss of >3% severely impacts typing ability

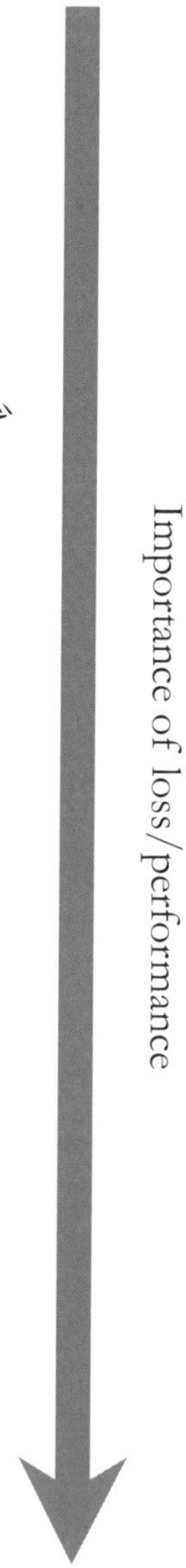

TCP bandwidth < (MSS/RTT)*(1/sqrt(loss))

Predicted throughput versus ttcp measured

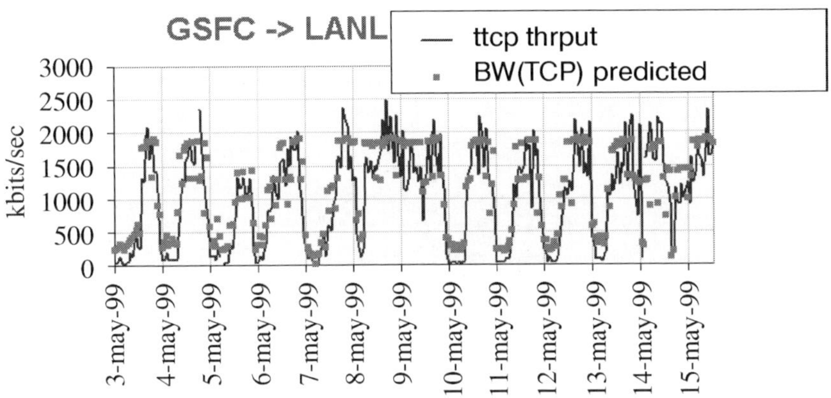

Measured TCP through put vs. predicted for non-zero packet loss

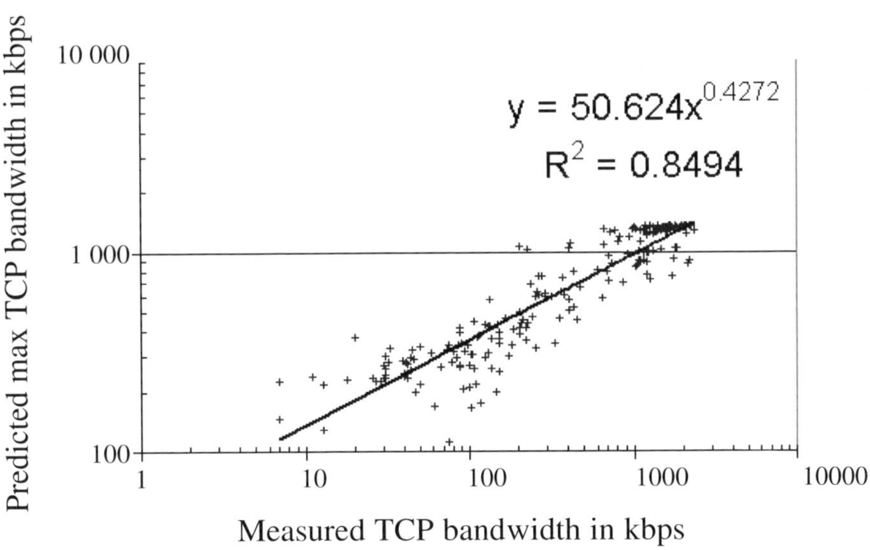

Relates to Web performance

(small files dominated by RTT)

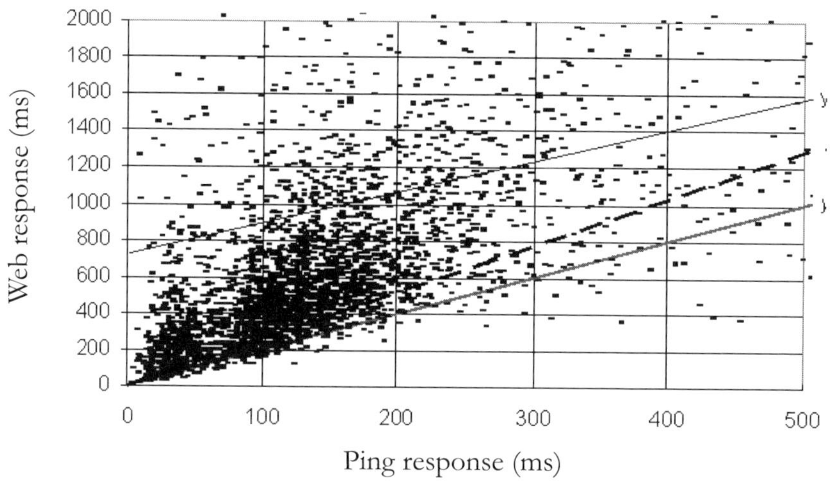

Residual = GET - 2 * min (ping RTT)

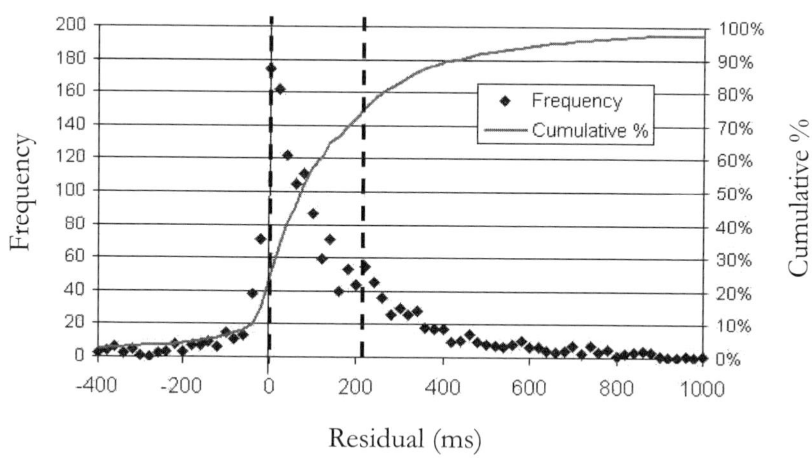

Bulk transfer — performance trends

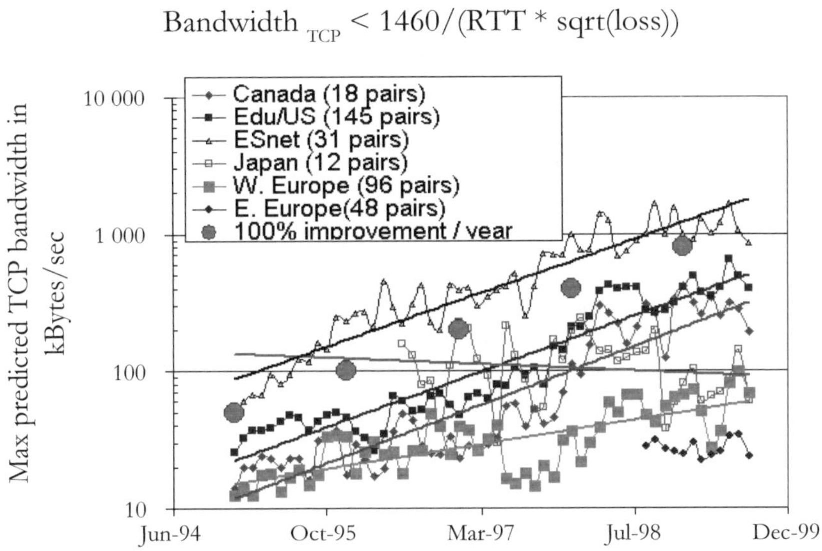

Interactive apps — delay

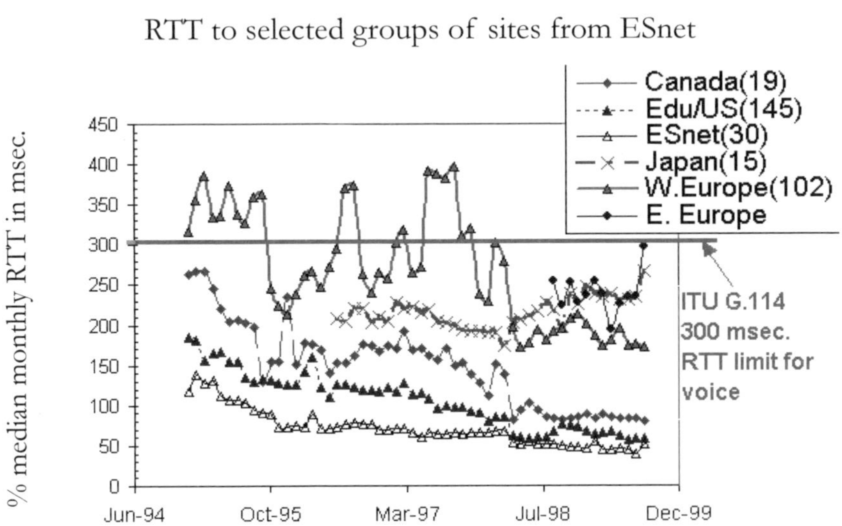

Interactive apps — packet loss

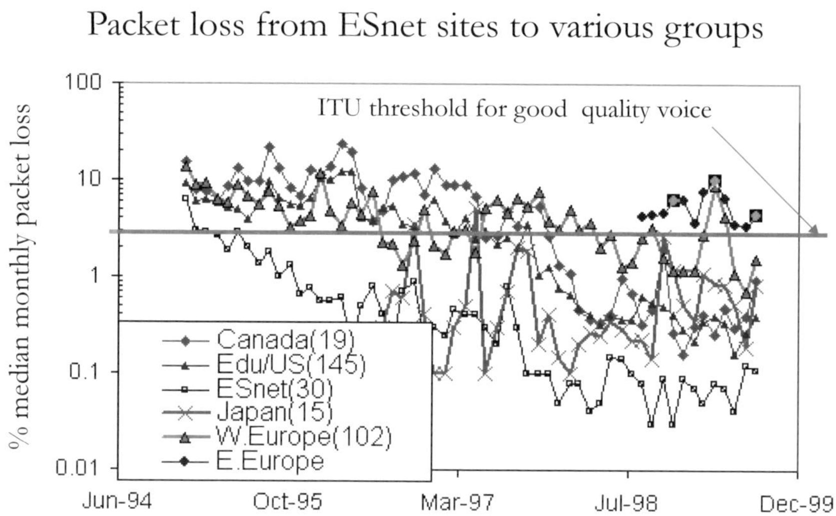

Interactive apps — Jitter

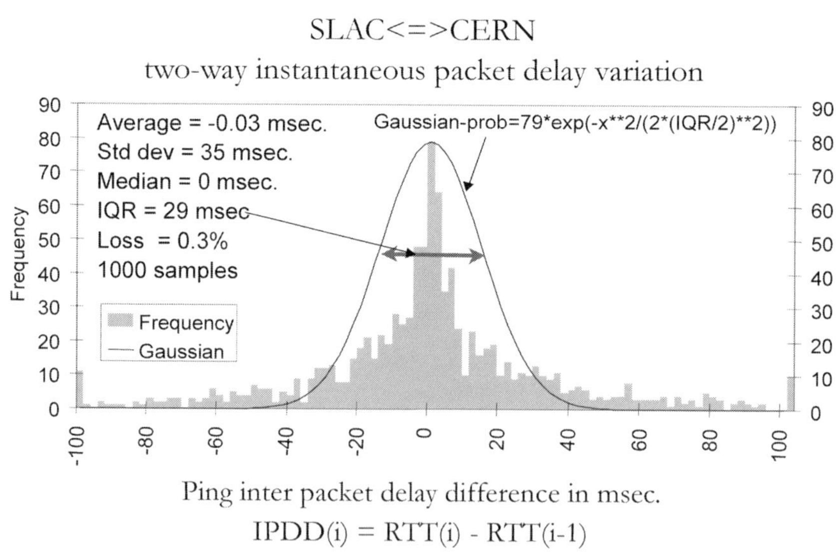

SLAC-CERN — Jitter

The statistics were last updated **Tuesday, 15 December 1998 at 19:21** ,
at which time **'Cernh8.cern.ch'** had been up for **167 days, 0:54:05**.

`Daily' Graph (5 Minute Average)

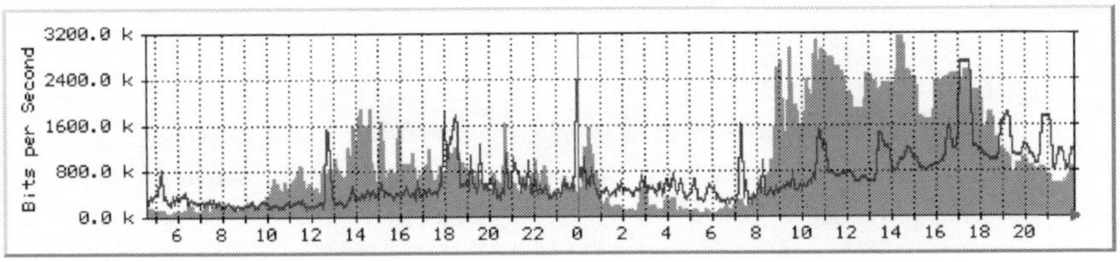

Max In: 3244.6 kb/s (82.1%) Average In: 1354.4 kb/s (34.3%) Current In: 1249.5 kb/s (31.6%)
Max Out: 2692.7 kb/s (68.1%) Average Out: 997.7 kb/s (25.2%) Current Out: 1307.9 kb/s (33.1%)

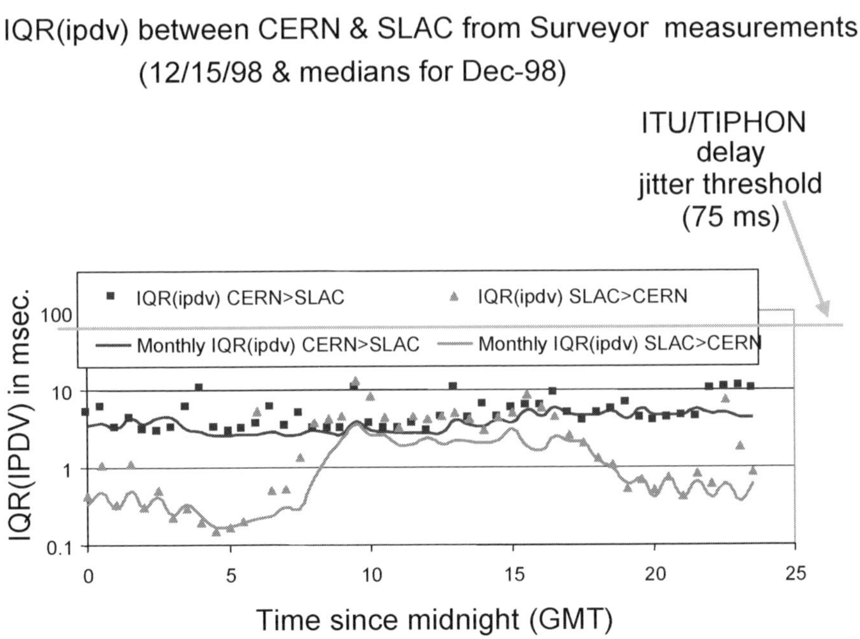

Availability — routing convergence

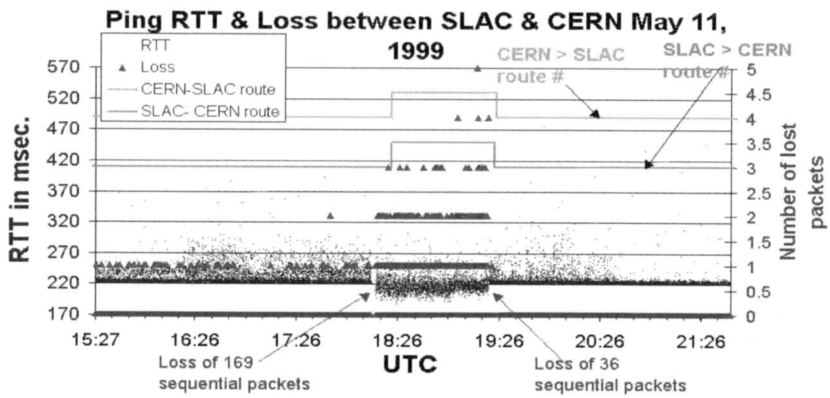

Availability — outage probability

- Surveyor probes randomly 2/seconds;
- Measure time (*Outage length*) consecutive probes don't get through.

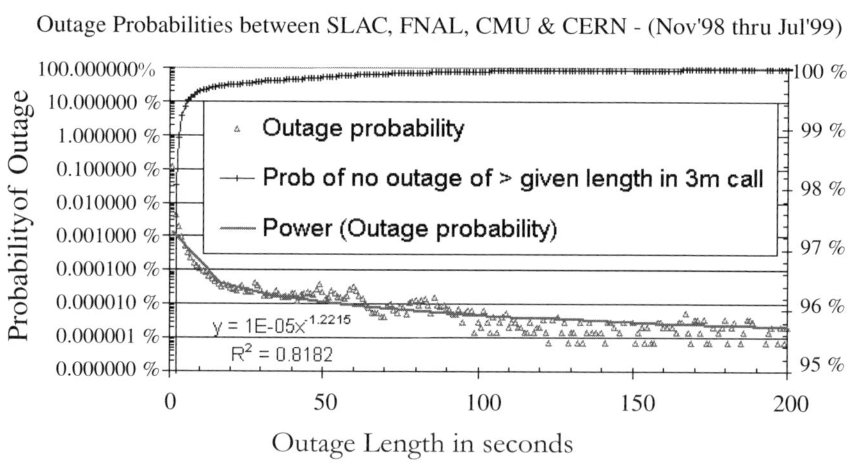

Error free seconds

- Typical US phone company objectives are 99.6–99.99%
- What do we see for the Internet using Surveyor measurements.

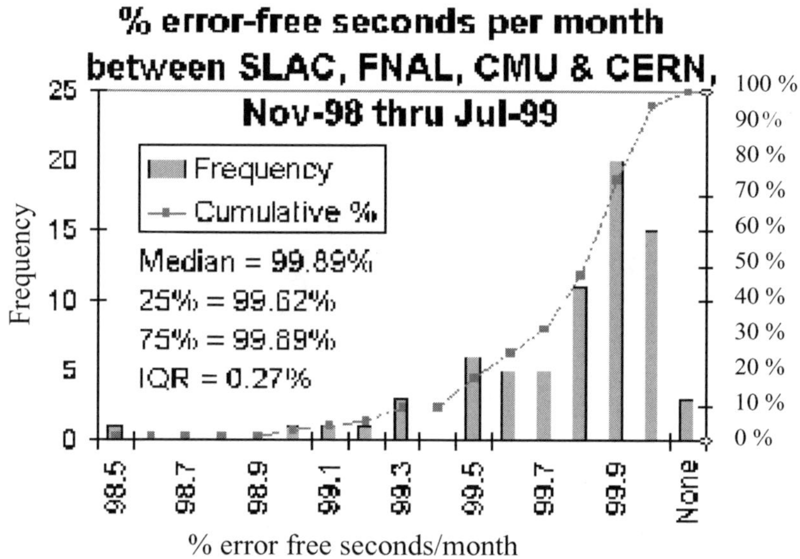

Improving Quality of Service (QoS)

- More bandwidth
  - Keep network load low (< 30%),
  - Costs (at least in the Web) are coming down dramatically;
- Reserved/managed bandwidth
  - generally on ATM via PVCs today;
- Differentiated services.

More bandwidth

Packet loss between ESnet & UK since 1995

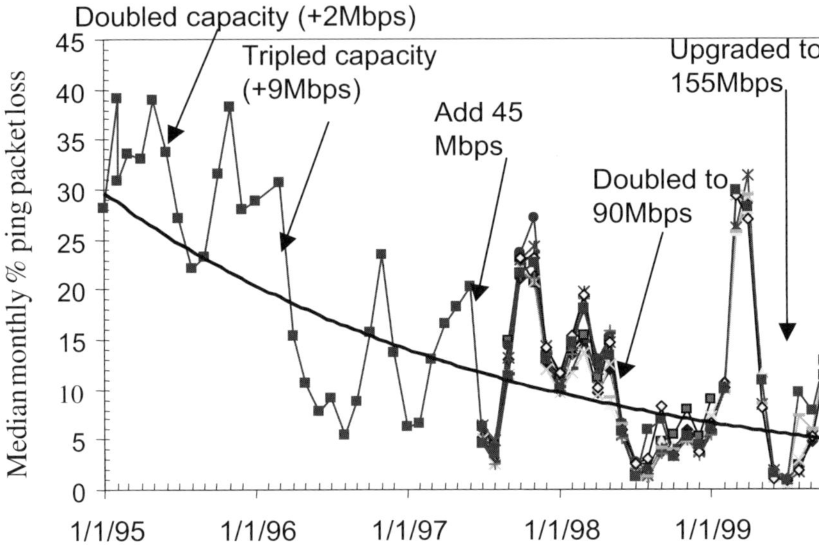

- Holidays also have dips;
- Transatlantic bandwidth is quickly absorbed;
- Jan-95 had 2 Mbps, now at 2×OC3 so 150 times increase in bandwidth over 4.5 years.

Reserved bandwidth

- U.K. transatlantic link at 2×155 Mbps, will reserve 2% for special projects both short & long term;
- CERN & Italy both have reserved bandwidth to US;

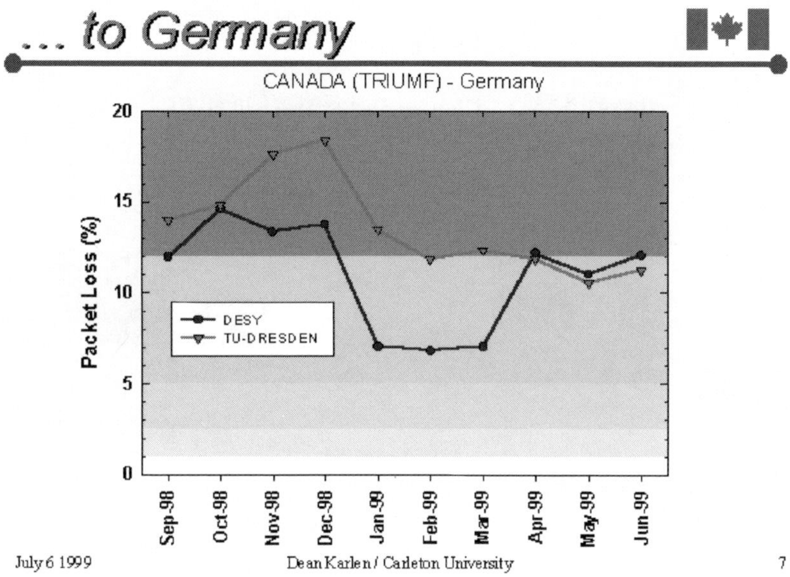

- DESY had reserved bandwidth to ESnet, but not to

  – North America in general, so:
  – performance to Canada & .edu bad
  – performance to ESnet good to acceptable.

Reserved BW — DESY & .edu/.ca

- DESY worked with DFN to provide 3.5 Mbps (< 3% total) non-shared bandwidth (PVC) for DESY to major educational sites in North America starting August 12, 1999;

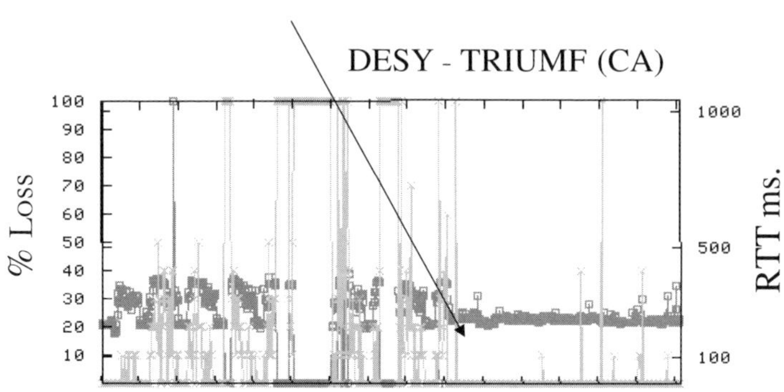

- Rest of Germany still around 12% loss (vs 1–2%).

Interactive apps — VoIP

- Pilot VoIP between 5 national Labs:
  - CERN (Geneva), DESY (Hamburg), FNAL (Chicago), SLAC (San Francisco), LBNL (Berkeley).

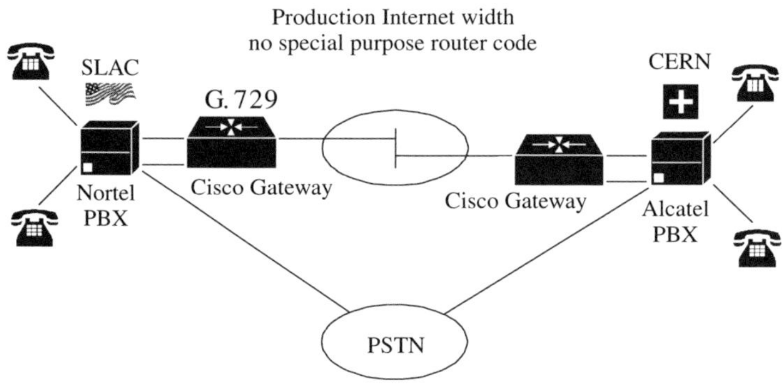

Differentiated services

- Provides improved performance for small fraction of traffic;
- Quite complex, requires policy, reservations, signaling, classification/marking, metering, policing (shaping & dropping), queuing/scheduling (congestion management), cross AS agreements
  - still research & pilots;
- SLAC & LBNL have a DS testbed with a 3.5 Mbps ATM PVC carved out of 43 Mbps
  - Make phone call,
  - No load phone call is fine,
  - Inject 4 Mbps UDP load;

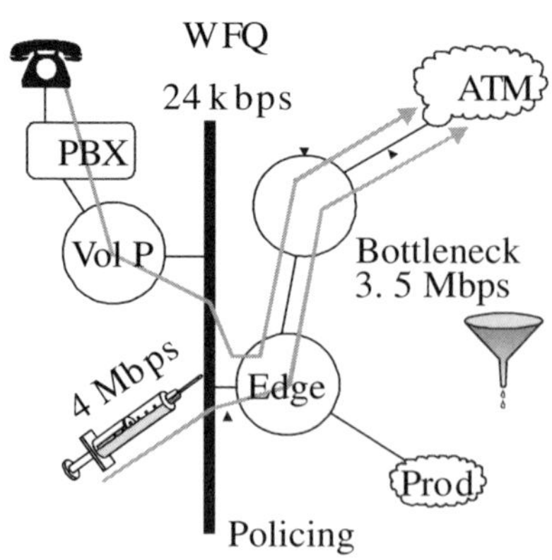

- No WFQ can't make call
  - If make call then terrible quality;
- Apply WFQ & policing (via CAR),
- With WFW call sounds fine
  - Next use ping to characterize:
- Mark ping TOS bits with CAR, & use WFQ in routers and see how it affects loss, RTT, jitter etc.

Differentiated services 2/2

- Pilot at ANL
  - 50 Mbps reserved for pilot network,
  - UDP traffic generator sending at 80 Mbps,
  - Cisco CAR for policing & WFQ for prioritizing,

Video/audio performance	No diff serv	With diff serv
No load	Good	Good
80 Mbps load	V. Bad	Good

  - Reservation manager signaled DS routers to set up & tear down reservation.

Conclusions

- Performance is getting better;
- Within Western R&E networks things are good
  - Good enough even for VoIP in terms of RTT, jitter, loss;
- But keeping pace takes constant upgrades;
- Transoceanic, needs special care;
- E. Europe, Russia, China, S. America performance is where N. America & W. Europe were 4 years ago;
- Peering is critical;
- Internet reliability, even in the West, has a way to go to meet phone company standards of 99.999%

More information

- **IEPM/PingER home site**
  - http://www-iepm.slac.stanford.edu/;

- **Surveyor/IETF/IPPM project**
  - http://www.advanced.org/csg-ippm/;
- **ICFA-SCIC Homepage**
  - http://www.hep.net/ICFA/index.html.

ICFA-SCIC charge

- Make recommendations to ICFA concerning the connectivity between America, Asia and Europe;
  - As part of the process of developing these recommendations, the committee should:
    - monitor traffic,
    - keep track of technology developments,
    - periodically review forecasts of future bandwidth needs, and
- provide early warning of potential problems;
- Create subcommittees when necessary to meet the charge;
- The chair of the committee should report to ICFA once a year, at its joint meeting with laboratory directors.

Comparisons of networks

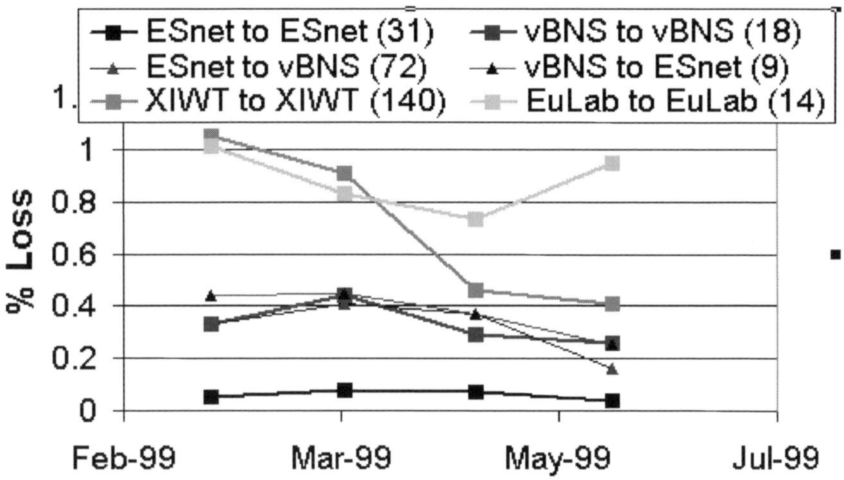

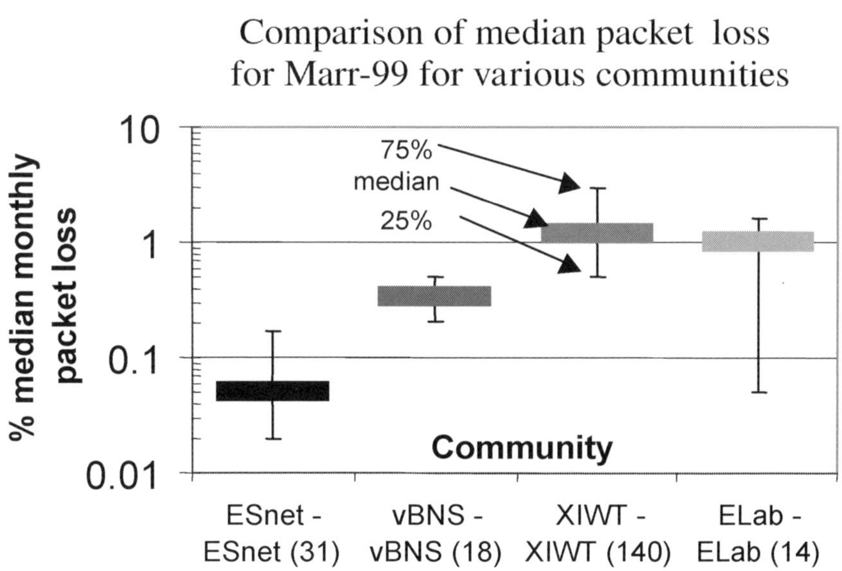

Few islands of good performance

From Country To Continent Packet Loss Sep 99 Submit

WORLD	Brazil	Canada	Denmark	Germany	Hungary	Italy	Japan	Russia	South Korea	Switzerland	United Kingdom	United States
South Pacific	10.94	7.16	5.69	6.16	10.23	8.64	3.17	27.04	7.45	1.41	2.54	5.24
West Europe	8.50	3.55	1.05	2.44	3.14	0.47	10.43	10.41	6.42	1.27	0.89	2.71
East Europe	19.01	14.18	4.82	2.77	6.09	3.30	14.61	12.11	16.66	3.42	3.04	8.19
North America	8.24	0.82	1.61	1.96	7.82	4.77	3.24	28.87	1.11	0.59	3.08	0.29
Asia	8.55	3.23	10.31	3.68	8.46	9.80	0.43	17.55	4.01	2.59	7.48	2.18
South America	1.94	16.10		14.33	19.40	24.66	20.16	34.12	19.61	8.33		15.38
.	Brazil	Canada	Denmark	Germany	Hungary	Italy	Japan	Russia	South Korea	Switzerland	United Kingdom	United States

Performance between PingER TLDs

WORLD	UK	IT	NET	DE	GOV	JP	CH	HU	EDU	CA	DK
UK	0.07	2.00	1.72	1.60	1.77	5.88	1.66	2.21	1.37	1.71	1.14
IT	1.42	0.94	0.55	1.12	1.65	5.00	0.86	2.09	1.03	3.75	1.15
NET	1.74	3.14	3.05	0.85	3.27	2.43	2.95	3.32	2.27	3.91	2.17
DE	1.58	1.63	0.16	0.26	0.84	8.14	0.62	2.24	3.90	10.62	1.31
GOV	1.63	1.07	0.03	0.35	0.20	2.10	0.61	4.16	0.09	2.74	4.03
JP	6.89	8.57	0.66	7.92	1.45	0.12	4.95	8.27	0.90	1.34	7.88
CH	1.21	1.15	0.51	0.59	1.57	5.58	0.31	1.57	0.51	2.76	0.50
HU	6.74	2.95	4.02	2.19	4.73	12.66	3.99	1.80	5.51	5.20	6.61
EDU	1.57	2.85	0.16	4.10	1.30	1.49	0.87	3.37	0.18	1.44	1.73
CA	1.60	4.67	0.23	4.82	1.60	1.08	0.70	4.15	0.52	1.33	2.42
DK	0.98	1.74	2.85	1.27	3.07	4.65	0.64	1.69	2.52	2.35	

Non-monitoring sites

CZ	1.80	1.80	5.08	1.42	5.44	6.72	1.40	2.15	4.90	5.78	1.52
ES	2.41	3.56	12.53	3.43	14.27	10.78	2.60	3.82	11.09	12.32	2.37
SU	3.18	6.11	1.44	1.31	0.89	4.25	2.59	4.20	10.91	16.23	3.89
AU	3.83	4.03	3.51	5.32	4.43	2.04	3.29	3.93	3.76	3.09	3.82
FR	1.35	1.36	0.62	1.06	1.53	5.06	0.38	2.30	0.44	2.05	0.75
CN	7.94	9.55	1.69	9.27	2.74	1.66	7.98	9.75	2.80	3.96	9.54
TW	7.04	3.28	1.79	3.75	2.50	4.07	1.35	3.88	0.36	3.99	9.73
RU	4.00	5.40	2.68	2.29	12.16	19.69	3.12	7.27	24.12	29.52	21.75
NZ	8.21	11.89	6.09	12.44	6.30	11.12		10.62	3.21	40.24	
SE		1.81	1.98	2.48	2.27	6.58	0.63	2.93	3.25	3.62	
AR		14.39	13.45	13.42	14.90	13.66	12.41	19.43	13.26	13.75	
CO		14.21	7.05	15.32	11.31	9.96	5.00	25.53	10.62	12.51	
BR			2.91	1.90	3.00	2.24	2.33	5.79	3.54	4.19	
IN			2.46	2.97	2.91	7.81	2.91	8.69	2.50	3.53	
NL	1.07	1.85	0.11	0.47			0.78	1.39	0.12		
ORG	3.54		2.94		4.40		1.29		0.06		
COM			0.29	14.41		8.96			2.26	20.00	
BE			1.30				1.14		1.59		
PL			24.40				9.41				
IL							1.42		8.14		
IE			0.97				2.32				
KR						2.77			0.74		
NO									3.82		
GE									0.54		
US									3.12		
SI									8.20		
MIL					5.32						
:	UK	IT	NET	DE	GOV	JP	CH	HU	EDU	CA	DK

Outage frequency between SLAC, FNAL, CMU & CERN (Nov'98 thru Jul'99)

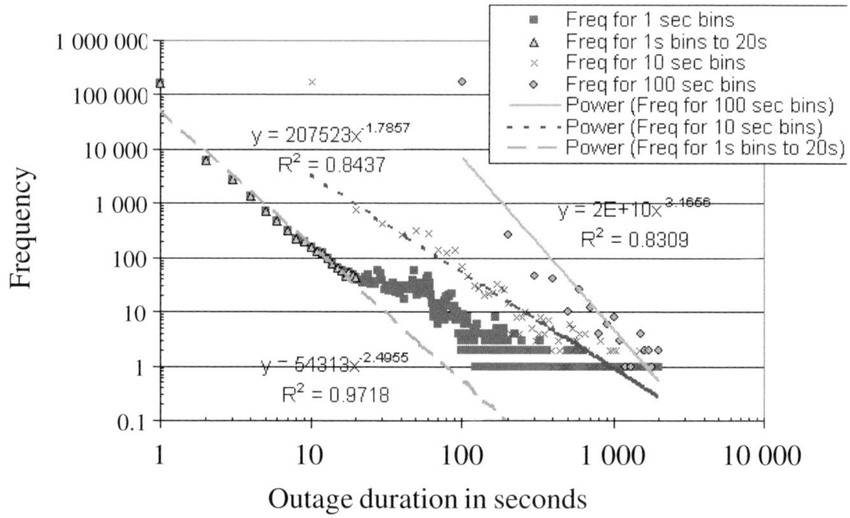

Interactive apps — availability

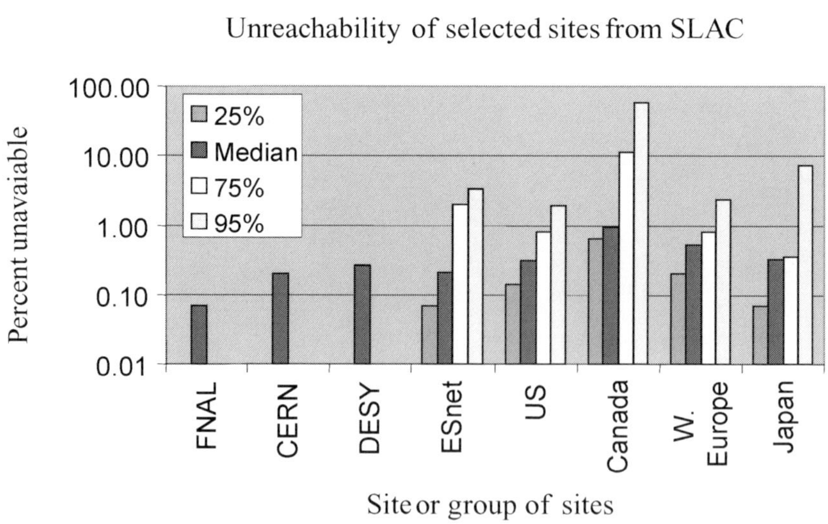

ESnet QoS testbed

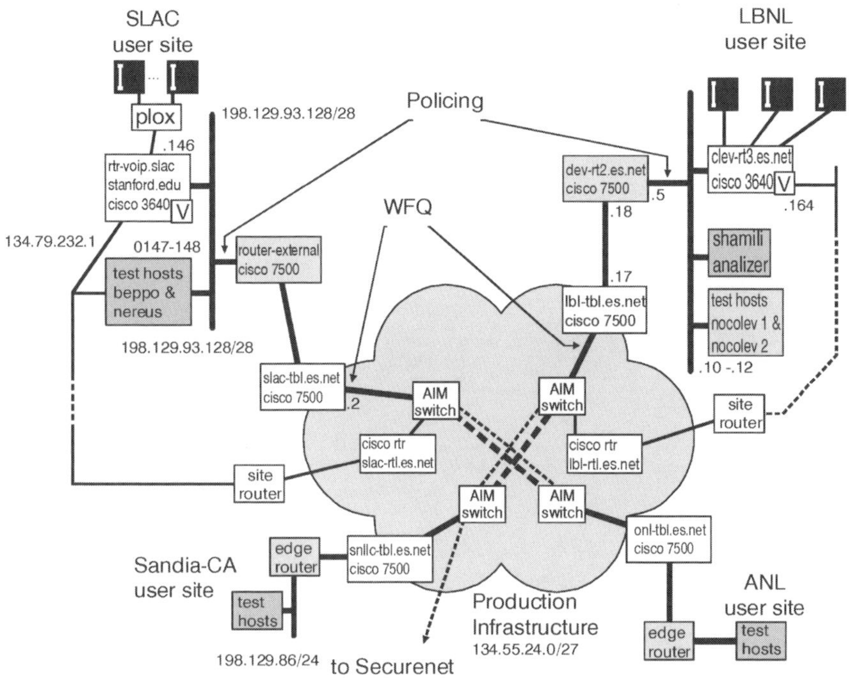

Work in progress

- Redouble efforts to get HEPNRC more involved;
- Random scheduling of pings (in beta at 2 sites);
- Recording more information (in production at XIWT);
- More flexibility in choice of packet sizes, number of packets in sample, frequencies (tailor to bandwidth and performance between pair);
- Look for ICMP rate limiting signatures;
- Gather historical traceroute information for PingER;
- Calibrate ping jitter against VoIP jitter;
- Passive RTT using SYN/ACK, FIN & PSH;
- Calibrate using ping to measure QoS effects

  – setting up QoS testbed between SLAC & LBNL;

- Measure frequency of outage lenths;
- Continue work with XIWT, Surveyor, AMP & RIPE;
- Other possibilities:

- Thinking about extending framework to other "apps", e.g. following IPMP work, TCP/UDP echo, http (CERN interested, XIWT also interested, there are also commercial tools, but expensive),
- Generate alerts (HEPNRC).

Networking Developments in the Caucasus Region
J.-P. Nadreau and O.B. Popov (Eds.)
IOS Press, 2001

Sergey Berezhnev

Organization of Telecommunication System for Science Community on the Base of Broadcast Web-cache Technology as a Further Development of RADIO-MSU Network

Broadcast Web-cache technology is a combination of a system satellite broadcasting data transmission analogous to IP-Advantage Hughes technologies and Web-cache technology, but with the integration of their advantages.

Web-cache technology

Web-cache technology is used to raise the standards of service provided to the clients using the Web and to economize the capacity of backbones of the given node. The principle of operation of Web-cache technology is based on the idea that any document required by a Web user is kept at a special server (cache-server). Later on, if the same document is required by another user, it is provided from the cache server, which prevents from being repeated providing document via the backbone. Thus the backbone capacity is economized. Besides, a document from cache-server can be received by a system user much faster than it was sent from the original server. So, the main purposes of the using of Web-cache technologies are the following:

- Economizing backbone capacity;
- Raising the quality of services provided to the clients.

It should be noted that the efficiency of the cache-server operation is directly dependent on the disk space volume, where Web documents are kept and on the number of the requests (or users) which it processes:

- The greater the number of documents kept on cache-server, the greater the possibility that another required document will be found on cache-server and not required via backbones;
- The more requests processed by cache-server, the more documents kept on it locally, and further transmission from original Web-servers is not require.

The practice of using cache-server by RADIO-MSU demonstrates that when the disk space volume is 20 GB, the percentage of providing Web-documents from cache-server when users are working with Internet.

Broadcast Web-cache technology

The use of satellite communication links make it possible to develop the Web-cache technology idea and increase its efficiency immensely.

The key idea is to use one satellite channel which is received at all network nodes and via which all Web documents are transmitted. The documents are kept at each cache-server and the users of the given node have access to them according to cache-server technology, described above. If a requested document is not found at cache-server, then the request is sent to the central node via the usual channel. The document requested is received by the central cache-server and is transmitted to broadcasting satellite channel. Thus, the user receives the document requested. Besides, the document is kept at all network nodes and any user of these nodes will get it from a local cache-server, which is much faster and does not require transmission of the document via backbone.

The given technology not only makes it possible to economize communication channels capacity from the centre to each network node, but also to have one wide satellite communication link which is received at every node.

The existence of a single wide satellite channel instead of several minor channels of the same total capacity greatly economizes frequency and energy resources of the satellite. The existence of the usual two-way individual channel between each peripheral node and the centre node is necessary. The channel is used both to transmit Web ask to the central node as well as for the operation of the all rest protocols.

The basic distinction in realization of broadcast Web-cache technology, is associated with the variants of the organization of individual channels between the peripheral and the centre nodes:

- Satellite channels in the framework of the system;
- Leased lines;
- Tunnel via another ISP.

Besides, there is also a possibility of organizing a system whereby individual channels from the centre node to the peripherial nodes are included in a broadcasting satellite channel.

The specification of an individual channel from the general stream at a peripheral node is carried out by a peripheral router.

The advantages of including individual channels in a common broadcasting channel are the following:

- Economizing the resources of the satellite;
- A possibility of dynamic redistribution of channel capacity between various peripheral nodes;

- A possibility to use a smaller agregated capacity of the satellite than by using a separate channel for each node. This is guaranteed both by the possibility of a dynamic redistribution of capacity and the effect of static multiplexing of the traffic. The latter appears as the result of a natural time non-homogeneous work of users, as well as the time difference at various network nodes.

So, this is the means that we suggest be used in the future for further development of RADIO-MSU network.

Networking Developments in the Caucasus Region
J.-P. Nadreau and O.B. Popov (Eds.)
IOS Press, 2001

Peter Rastl

ACONet's Support to Networking in Central & Eastern Europe

Motivation for this presentation

- In the early nineties ACONet's support was instrumental in promoting research networking in Central and Eastern Europe;
- What can be learned from this experience for other regions ?

Historical situation

- Political changes in CEE countries
  - unrestricted access to new technologies,
  - new market for computer manufacturers;
- Weak national coordination
  - no NRNs, only initiatives from individual institutions;
- Market-place for missionaries
  - european Commission tried to boost OSI protocols;
- Internet was not yet recognized as an economical engine.

Why Austria?

- CEE headquarters of computer companies were located in Austria
  - manufacturers contracted support from Vienna University;
- Support from Austrian government
  - vice-chancellor Erhard Busek was Minister of Science;
- Cultural ties with neighbour countries

- Germany was busy with former German Democratic Republic;

- Austrian network was not too developed
  - Austrian technological support was not considered a threat.

What has been done?

- Education of networking experts. "Network Seminars for Scientists and Network Managers from Central and Eastern Europe", sponsored by DEC:
  - 5 seminars with 283 participants from 16 countries held in 1991–1993;
- Computer systems for universities. IBM Central and Eastern Europe Academic Initiative (1991–1994) contracted Vienna University for consultancy and education:
  - Implementation of infrastructure in 10 countries;
- Platform for exchange of experiences
  - ACONet organized several coordination meetings in 1992 and 1993 that led to the creation of the CEENet Association (15 Jan 1994);
- Subsidy for Internet access in Vienna
  - Austrian leased line costs for 64 kbps link to Ebone node in Vienna.

What were the results?

- Creation of national research networks
  - concentration and coordination of national efforts make NRN more attractive to sponsors;
- Acquisition and exchange of experience
  - NRN management and staff is able to cope with technical and organizational issues and participates in international projects;
- Initiating awareness, developing funding sources
  - networks are generally accepted as a mandatory infrastructure which has to be funded from the research budget;
- Networking expertise superior to others
  - NRN catalyzes the commercial Internet market. NRN offers national coordination services (e.g. DNS administration, Internet exchange).

The lesson?

- Warning: Conditions are different!

  – not the same time, not the same region, not the same players;

- Create NRNs

  – join forces, don't rival nationally, cooperate internationally;

- Offer services to a broad user community

  – regular use will create awareness and funding responsibilities;

- Don't try to be too innovative

  – users primarily need stable services, not only research projects;

- Develop NRN into a center of excellence

  – expertise and organizational stability will attract growth.

Networking Developments in the Caucasus Region
J.-P. Nadreau and O.B. Popov (Eds.)
IOS Press, 2001

Peter Villemoes

NORDUNet Works Well in the Nordic Countries

The Nordic Countries

Nordic networks for research and education

- National research networks:
  - Forsknings/Sektornet (Denmark),
  - FUNET (Finland),
  - INTIS (Iceland),
  - UNINETT (Norway),
  - SUNET (Sweden);
- About 500.000 IP nodes;
- Financed by public money (universities and ministries);
- Total budgets about 30 M $US plus NORDUNet costs.

NORDUNet history

Tradition for Nordic cooperation:

1980 — First NORDUNet conference on networking;
1985 — NORDUNet program funded by Nordic Council of Ministers to interconnect Nordic universities;
1985 — Nordic national networks begin;
1987 — Internet protocols selected;
1988 — Nordic national networks decide to cooperate on international connectivity;
1988 — NORDUNet connects Nordic NRNs to Europe and USA;
1993 — NORDUNet A/S established as legal entity.

NORDUNet mission

- International service provider for the Nordic NRNs, by connecting Nordic NRNs to each other and to
  - the rest of Europe,
  - The USA,
  - CEE countries;
- Common Nordic platform for international network collaboration with
  - Europe — QUANTUM, DANTE,
  - North America — UCAID (Internet2), NSF, Canarie,
  - Asia Pacific — APAN, NACSIS;
- International service provider for research networks in the Baltic states, Poland, Russia and Ukraine;
- Network development in the NORDUNet2 project.

NORDUNet2 project

- 3 year Nordic Council of Ministers project, 4 M $US;
- Catalyst for joining other Nordic programs;
- Platform for international collaboration (EU, USA, . . .);
- Participation from research, education, and industry
  - four pillars,
  - distance education,
  - telemedicine,
  - digital libraries;
- Middleware;
- Use existing NORDUNet network;
- Secretariat with project leader Anders Gillner in Oslo.

NORDUNet organisation

- NORDUNet A/S shareholders are the Nordic national research networks or their ministries;
- Board made up of the national network managers;
- Financed by the Nordic national networks;
- 20 M $US turnover in 1999;
- Lean, lightweight organisation;
- Outsource services to Nordic networks and universities:
  - operations (KTH in Stockholm),
  - technical management (RUNIT in Trondheim),
  - information (Tampere University),
  - security (UNI-C Copenhagen).

NORDUNet links

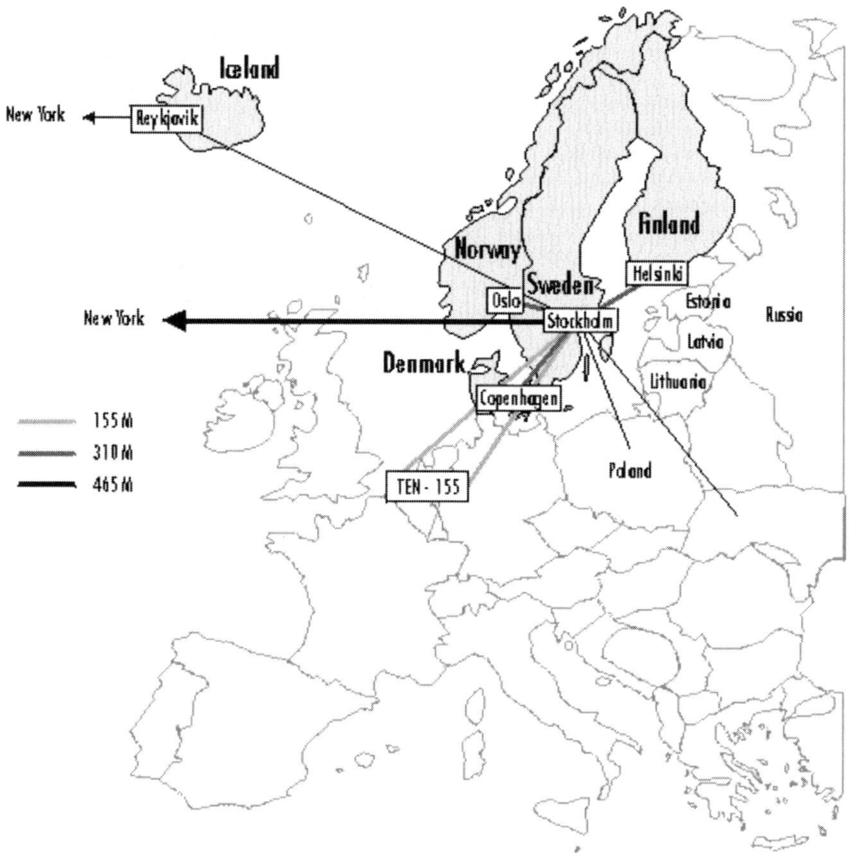

NORDUNet–USA connection

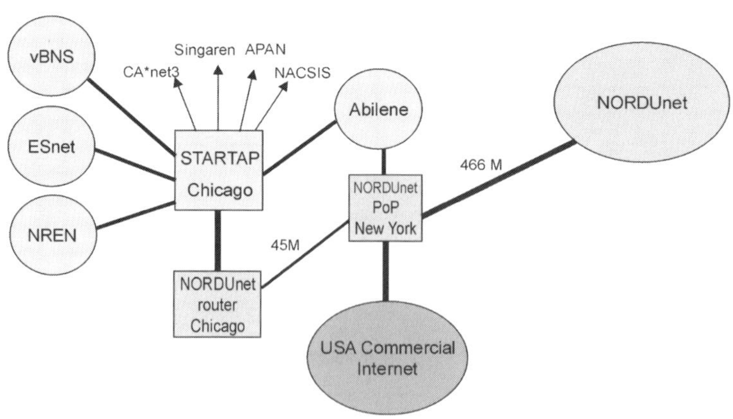

NORDUNet benefits

- Economy of scale
  - common access to TEN-155,
  - common bandwidth to the USA;
- Combined buying power invites good offers;
- Sharing resources between countries
  - specialists,
  - file servers;
- Unified platform for international collaboration
  - counts as a large European network country,
  - accepted as partner in North America and Asia Pacific research networking.

Preconditions for collaboration

- Must add value
  - shared resources,
  - buying power;
- Should have minimum overhead
  - distribution from shared links,
  - administration;
- Stable cost-sharing agreement
  - must allow growth,
  - GNP based,
  - usage based;
- Direct control by owners
  - service, not competition;
- Common network architecture
  - started multiprotocol, now IP only,
  - still ATM discussions;
- Equal stages of network development;
- Common views on network development.

Networking Developments in the Caucasus Region
J.-P. Nadreau and O.B. Popov (Eds.)
IOS Press, 2001

Walter Kaffenberger

Funding Possibilities through the NATO Science Programme

NATO's three dimensions

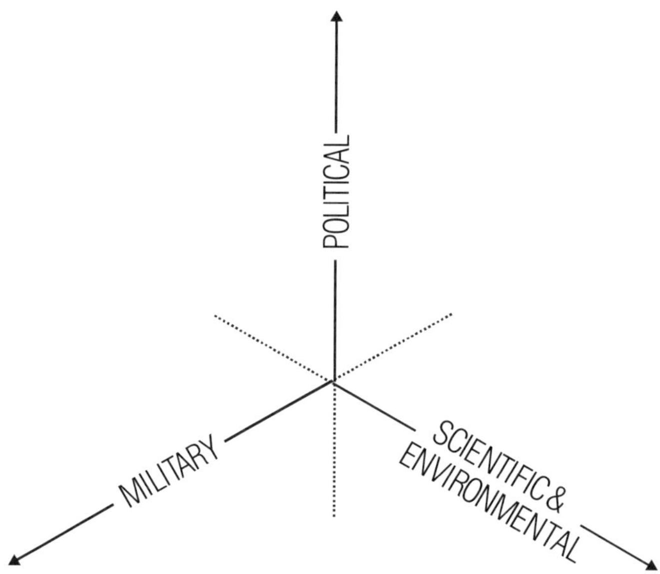

NATO science programme

- **The basic idea**:
  "Bringing scientists together for progress (in science) and peace."

Proportion of funds for projects with partners

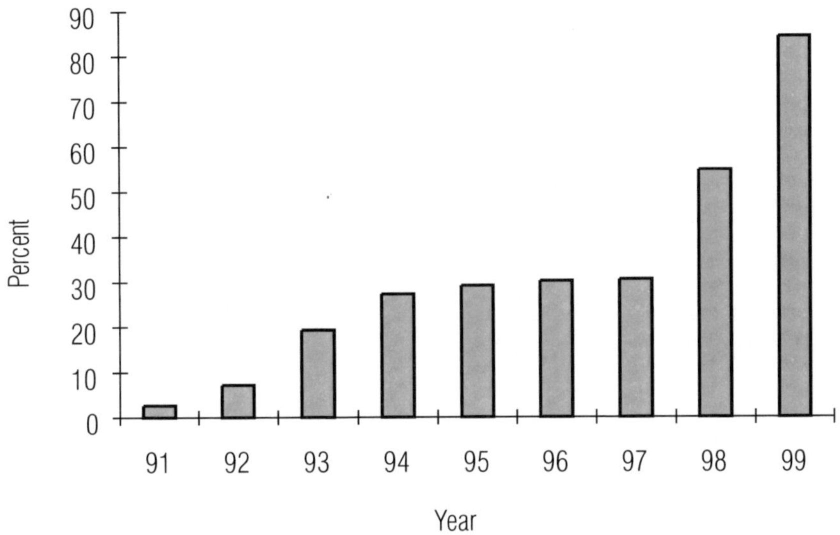

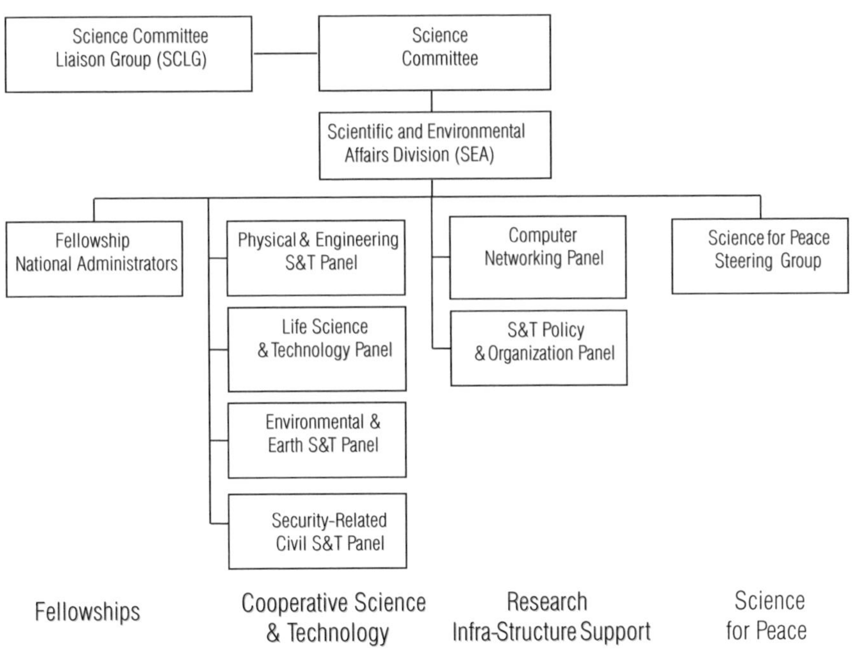

Collaborative science and technology

- **Support mechanisms**
  - advanced research workshop,
  - advanced study institute ('Summer School'),
  - collaborative linkage grant,
  - expert visit.

Area on computer networking

- **Goals**:
  - enhancement of telecommunication infrastructure,
  - development of computer networking policy.

Integration into scientific community

Area on computer networking

- **Support mechanisms**:
  - consultants,
  - advanced networking workshops,
  - networking infrastructure grants.

Computer networking consultants

- **Feasibility consultants**
  - help to define requirements for good proposals;

- **Infrastructure consultants**
  - support to spend award properly,
  - and help to setup the equipment.

Advanced networking workshop

- **Policy workshop**
  - increase harmony among network policies;
- **Training workshop**
  - transfer of technical know-how to network administrators.

Area on computer networking

Award for Training (in Kiev) of Network Administrators
to: Tugrul Yilmaz (tugrul-yilmaz@metu.edu.tr) and Slava Shkarupin (slava@kar.net)

NATO science programme

- **Summary of characteristics**
  - bottom-up program with little bureaucracy,
  - simple application forms,
  - concise proposals,
  - quick decisions,
  - high success rate.

Networking Developments in the Caucasus Region
J.-P. Nadreau and O.B. Popov (Eds.)
IOS Press, 2001

Juergen Hess

Open Society Institute

Open Society Institute

- Part of the Soros foundations network
- Budget $574.7 million (SFN) in 1998
- Coordinate regional programs
- Budapest and New York

OSI's Internet program

- Fundamental goals of the Internet program
  - to provide the free flow of information,
  - to promote an active and well-informed citizenry;
- Active from 1993 to 2000 (?);
- From 1994–1997 "survival connectivity";
- Training;
- Content development;
- Target: educational institutions, libraries, museums, hospitals, NGOs, and other non-profits;
- 750,000 — 1 million connections worldwide;
- 1998 — $8.8 million;
- 1999 — $4 million;
- Co-ordinate with other organizations, i.e. governments, international organizations (NATO, UNDP, IREX);
- Support from corporations (discount contracts with major computer companies);
- 1998 — $3 million;
- Foundations-related (education, training, content);
- Human rights and independent media;
- Internet policy (policy and training);
- Country-specific Internet programs are run by the Soros foundations locally (Georgia, Azerbaijan and Armenia);
- Regional Internet programs are run by OSI New York;
- http://www.soros.org/internet;
- Regional program (New York)

- program should be a pilot project,
- it should be self-sufficient in the near future,
- it should have other co-sponsors,
- it has to be below $100, 000.

OSI's Internet program in the Caucasus region

- Open Society Georgia Foundation;
- Open Society Institute — Azerbaijan;
- Open Society Institute Assistance Foundation — Armenia.

OSI's Internet program

- The Future
  - probably there will be a New York-coordinated Internet program for Central Asia and the Caucasus Region,
  - as before, we WANT to work together with other foundations, governments, and international organizations.

Networking Developments in the Caucasus Region
J.-P. Nadreau and O.B. Popov (Eds.)
IOS Press, 2001

Sandro Karumidze

Open Society — Georgia Foundation (OSGF) Internet Program

Started in 1995

Grant giving program in 1995–1996

E-mail / Internet Starter Kits;

Connectivity grants:
PC workstations, modems and Internet account;

Creation of Internet sites — homepages.

1996–1999 Support of Academy of Sciences & HEPI project with salaries.

The program began operating in 1997 with Internet Center for a Wide Open World — WOW

Funding

- 1995 — $150,000 — E-mail starter kit project;
- 1996 — $180,000 — Internet connectivity grants and web sites;
- 1997 — $360,000 — WOW, connectivity via phone in regions, content oriented projects, LAN for educational institutions;
- 1998 — $360,000 — WOW, content oriented projects, LANs, coding standards;
- 1999 — $270,000 WOW, content oriented projects, LANs, coding standards.

OSGF — programs and priorities

Education

- Higher education — state and private universities
- Secondary and primary education

Research

Distance learning within Georgia and abroad
Computer–Information society Education

- Libraries;
- NGOs;
- Legal reform;
- Medical institutions;
- Cultural heritage.

OSGF — partner organization

Donors

- USIS;
- Eurasia Foundation;
- World Bank;
- UNDP;
- Georgian government.

Partners

- NGOs — ITIC, NILC;
- IREX;
- Ministries of Education;
- Universities;
- Academy of Sciences;
- Operation WOW activities;
- Internet services
  - dial-up — NGOs, schools, state institutions — 250 institutions,
  - leased line — six universities, state library,
  - point of presence in Kutaisi — second largest city,
  - hosting of various web projects;
- Internet Public Access Site
  - open 78 hours per week — more than 170 people/week — more than 2000;
- Internet training
  - basic browsing, search and e-mail — about 2000 during two years;
  - web authoring — 200 people,
  - starting system administration training;
- Other projects
  - think quest,
  - EBSCO — Journals via Internet.

WOW today

- Staff of six people;
- Connectivity 384 kbps from a commercial ISP;
- 32 V.34 + dialup lines;
- Local traffic exchange with Academy of Sciences and High Energy Physics Institute — DESY Network;
- 12 workstations for visitors;
- Linux + Apache and NT 4.0 IIS 4 servers, Squid;
- Alkaline search engine etc.;
- Autonomous electric power system.

Grant giving of OSGF

- Public access sites;
- University LANs;
- Georgian content oriented projects — calls for proposals;
- The first Georgian web site contest in 1999;
- Support and promotion of a wide use of standards for coding of Georgian characters.

  $100,000 for grants in 1999.

Future plans

Cooperation with Universities and Academy of Sciences in creation of Georgian Research and Educational Networking Association — for the grant from NATO, Ramaz Kvatadze, the Executive Director, will talk about GRENA.

- Internet in Georgia — 1999;
- Population of Georgia 5.4 million.

Online Internet services started in 1995;
Great growth of Internet dial-up accounts in 1997—1998;
There are about 5–6 k dial-up accounts;
3–3.5 Mbps international connectivity;
5 primary and 6 secondary providers;
RIPE hosts 951;
Internet available in Tbilisi and two largest cities;
The number of computers sold per month is 500–1000, of which 25–30% goes to private owners — almost all with modems:

- Good
  - liberalization of telecommunication and licensing,
  - last state companies Telecom and city CO network going to be sold soon, but no regulation of Internet yet;

- Problems
  - power supply, no Georgian script software, few system administrators,
  - bad education system, few businesses and poor people,
  - no narrow band and few broadband digital data services,
  - bad telecommunication infrastructure in regions.

Networking Developments in the Caucasus Region
J.-P. Nadreau and O.B. Popov (Eds.)
IOS Press, 2001

Jacob Kuykendall

Broadband Internet Access via Satellite

Internet past, present and future

- Internet was developed to provide a survivable network f or US Government communications;
- Major universities were initial primary users;
- User-friendly access software has aided rapid migration to commercial and private sectors;
- Globally, businesses will not survive unless they use the Internet;
- Universities in developing and transitional countries must be leaders in preparing students for the Internet world;
- 500 M users, $1.3 T e-commerce by 2003.

Elements of a robust Internet infrastructure

- Bandwidth
  - domestic and university wide (LANS),
  - international bandwidth to reach peering nodes in US and Western Europe;
- Affordable local and long distance circuits;
- A management and technical staff to plan and operate the network;
- A government that supports development of the Internet infrastructure.

Limits to Internet growth in CEE

- Current low-speed access;
- High price of domestic circuits;
- High price of international circuits;
- Internet is a basic infrastructure element that is not understood by governments;
- Low priority relative to other budget items competing for limited funds.

Satellite Internet access

- Historically limited to 2 then 10 and finally 50 Mbps;

- Can now approach rates of 150 Mbps using BEM technology;
- Packet based bandwidth management allows secure Virtual Private Networks as a means of allocating the bandwidth;
- Single hop from up-link to user location In CEE, the lowest cost network solution.

Shared Internet access

- A regional satellite system allows cost-sharing with similar users;
- Users enjoy highly secure connections;
- **Each user can subscribe to bandwidth necessary for their requirements**;
- Multiple simultaneous transmission protocols;
- **Base load and over-subscription available**;
- Costs are related to delivered traffic.

Broadband satellite access

- Maximum data rate determined by transponder bandwidth and power;
- 32, 54 and 72 MHz bandwidths currently available
- Modern modulation formats support up to 155 Mbps
  - transmit power and receive antennas sized toprovide desired performance margins;
- Compatible with next generation Internet rates and formats.

CEESat.net access will provide Central and Eastern Europe a quantum advance

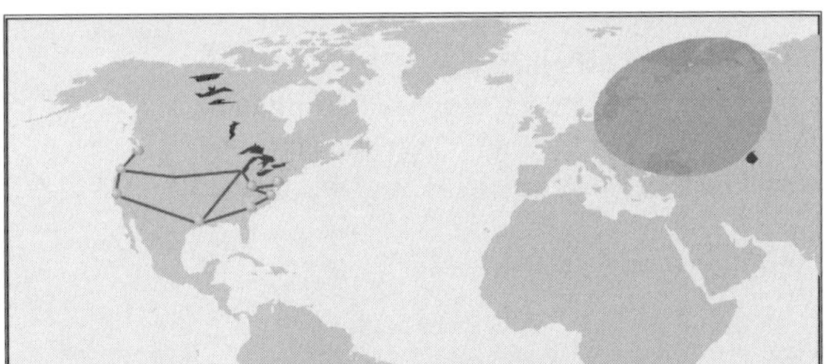

The next generation of Internet Backbone services for transitional nations

CEESat architecture

- Internet access to universities is CEESat priority

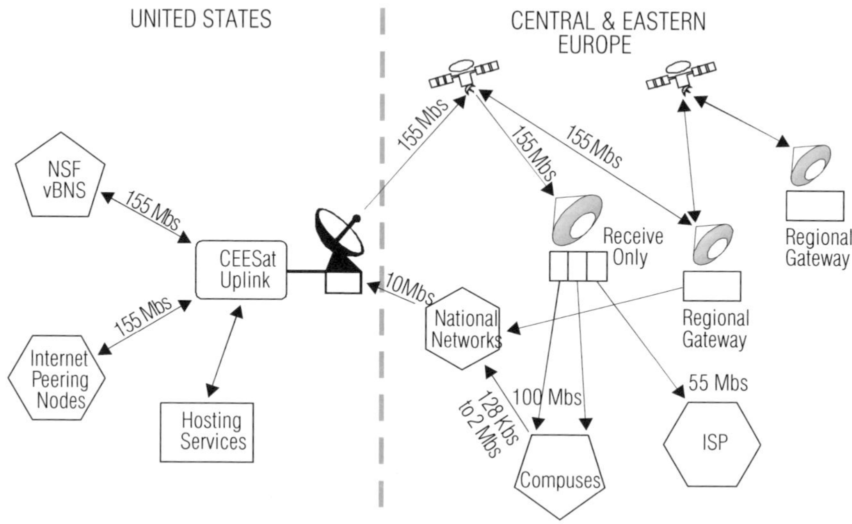

- Uplink hub to US Internet backbone with US & European satellite access nodes
  - TCP/IP protocol using ATM,
  - DVB and dedicated carrier protocols,
  - **All standard physical and datalink layer protocols supported**;
- Ethernet / IEEE / ISO / Token Ring / FDDI / SLIP / PPP / ISDN
  - StarTap international portal to NSF;
- Downlink terminals at user facilities;
- ATM / IP interface provides for local area distribution;
- Return links via national Internet or satellite gateways;
- Multiple backbone delivery protocols available including ATM / DVB and dedicated carrier;
- Oversubscription option available.

Benefits of CEESat

- Higher data rates;
- Can deliver bandwidth for peak demand on a dynamic basis;
- Readily available even in remote areas;
- Affordable;
- Compatible with future Internet protocols;
- Levels the playing field for small or emerging economies.

Bandwidth efficient modulation

From Lockheed Martin

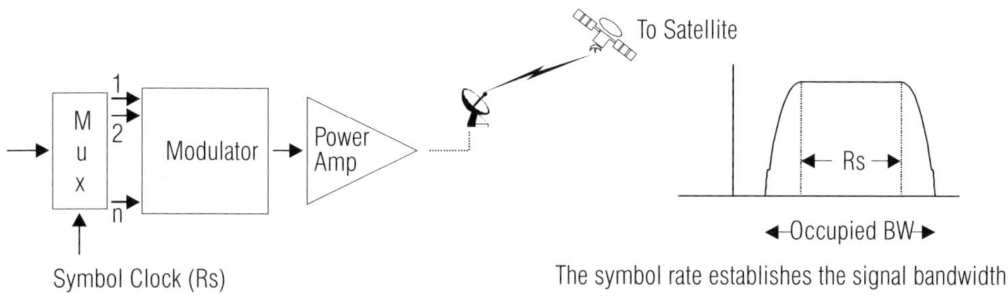

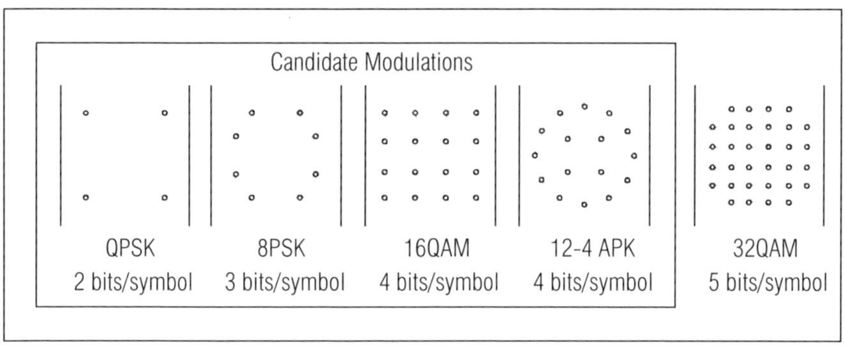

Increased bits per symbol increases the transmission data rate

Lockheed Martin

More than 40 years experience in government, commercial, and scientific satellite networks

Partial list of projects

- First communications satellite -1961;
- U.S. Satellite Defense Network;

- NATO III Satellite;
- 22 NATO III ground terminals;
- INMARSAT.

Global Internet - STARtap vBNS node

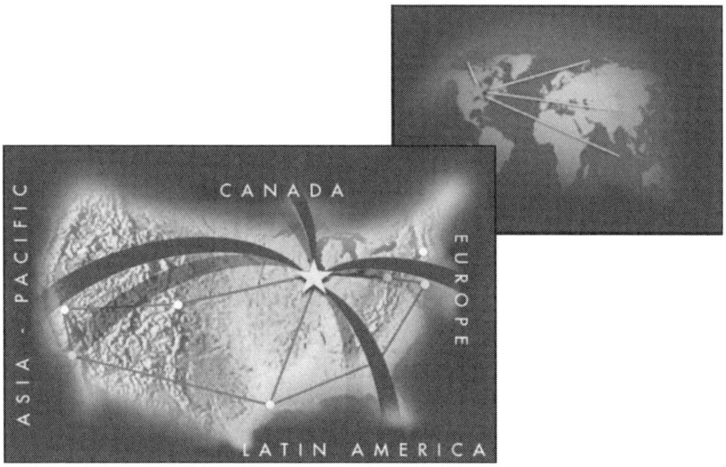

Summary

- Wide bandwidth Internet is the **key enabler** for economic viability;
- Satellites can provide wide bandwidth Internet to any locality now;
- Central and Eastern Europe can use **CEESat** to meet their present and future communication requirements;
- Contact:
  <jacob.kuykendall@gte.net>, fax: 1- 978-443-8636.

Networking Developments in the Caucasus Region
J.-P. Nadreau and O.B. Popov (Eds.)
IOS Press, 2001

Jan Gruntorád

Managerial and Organizational Methods for NRNs

Presentation overview

1. Usual start of academic networking;
2. Experience learned from the "West";
3. The CESNet story;
4. Recommendation.

Academic networking start

- Activity of a few enthusiastic individuals;
- Bottom-up approach;
- Get involved parties together;
- Get organized;
- Introduce managable structures.

Lessons learned from the "West"

- Special legal entity in charge with R&E networking created;
- Support (funding) provided from various governmental bodies;
- High-level supervisory board in place (strategic decisions, budget approval, etc.);
- Managing director in charge with day-to-day management (subcontracts of various tasks);
- Introduce user feedback (meetings, conferences, etc.).

Evolution of CESNet

1990 — EARN introduced to Czechoslovakia and run by Czech Technical University in Prague;
1991 — Internet connection over 9.6 kbps links to University of Linz, Austria;

1992 — Grant for building country-wide IP network for Czech Universities (20 MKbps — from the Czech Ministry of Education);

1993 — CESNet network (64 kbps backbone) in nine cities operational (run by Czech Technical University in Prague);

— CESNet Advisory Board nominated by the Czech Ministry of Education;

1994 — Services provided also to organizations other than R&E;

1995 — Czech Ministry of Education terminates funding of the CESNet network;

1996 — Creation of CESNet, Association of Legal Entities;

— Tender for implementation of the TEN-34 CZ network and pilot applications;

— Start of the TEN-34 CZ programme (1996–1998) — 550 Mbps (Government Board for Research and Development, Czech Ministry of Education);

1997 — TEN-34 CZ network (34 Mbps backbone) in nine cities operational;

— Start of migration of R&E users from CESNetS network to TEN-34 CZ;

1998 — Part of backbone upgraded to 155 Mbps;

— Migration of R&E users to TEN-34 CZ completed;

— CESNet network becomes a commercial ISP;

1999 — Contract with Czech Ministry of Education (Government Board for Research and Development) for 5 years (1999–2003) support of "High Speed Network for National Research and its New Applications";

— Decision to sell the CESNet network.

Recommendations

- Get organized, try to avoid competition between various interest groups, i.e. you can obtain much more together than individually;
- Approach appropriate governmental officials with well-prepared proposals;
- Use "good" examples from abroad;
- Explain idea of R&E networking to telecoms and other connectivity providers;
- Involve politicians.

Ramaz Kvatadze

Georgian Research and Educational Networking Association (GRENA)

GRENA was founded on July 26, 1999 by the following parties:

- Georgian Academy of Sciences;
- Tbilisi State University;
- Georgian State Technical University;
- Tbilisi State Medical University;
- Tbilisi State Pedagogical University;
- Georgian State Agrarian University;
- High Energy Physics Institute of TSU;
- Open Society — Georgian Foundation.

Goals of GRENA

- Development, construction and maintenance of the computer and communication infrastructure in Tbilisi to connect academic and research institutions to a unique information system;
- Providing connectivity of academic and research institutions in the regional cities — Kutaisi, Batumi, Poti, Rustavi to the Internet through Tbilisi;
- Connection with international networks;
- Development and construction of information nodes and networks.

Steps in the development of GRENA

- 1994 NATO / INTAS grant Georgia Academic & Research Network Project;
- 1997 NATO grant Improvement of the Georgian Research and Education Network capabilities;
- 1999 NATO grant development of Networking Infrastructure of Georgian Research and Educational Networking Association;
- OSGF — Equipment and Operational Expenses.

At present, more than 250 institutions are connected to the Internet through HEPI, ICM and the Internet Center WOW.

Connectivity

- HEPI — Satellite Station — bandwidth 128 kbps to DFN, before the end 1999 — 256 kbps;
- WOW — 384 kbps leased line from commercial provider SANet.

Basic services of GRENA

- Internet connectivity for educational and research institutions, libraries and other organizations through radio modems, leased lines and dial-up connections;
- Design and development of LAN for these institutions;
- Standard information services — WWW hosting, FTP, e-mail, etc.;
- User training and support.

Staff

- Administration (2–3 pers.);
- Technical department (10–12 pers.).

Networking Developments in the Caucasus Region
J.-P. Nadreau and O.B. Popov (Eds.)
IOS Press, 2001

Igor Mkrtumyan

Internet in Armenia

1. Introduction

The end of the 20th century is characterized by unprecedented achievements in telecommunication and Internet technologies. A virtual world created by the Internet provides new perspectives in education, work, commerce, and entertainment, representing an annual growth rate of 46 percent. The number of Europeans with Internet access is expected to surge to 43 million by 2003. There are now over 43.2 million Internet hosts.

The main usage of the Internet is in entertainment, education, e-commerce, and communications [1]. We will accordingly analyze the situation in Armenia in the education, e-commerce, and communications fields. Nowadays, the country in progress on the road to Information Society is characterized by such parameters as GDP per capita, market value, computers per capita, connections to Internet.

In Table 1, the figures for some of the countries as well as Armenia are shown. They are sorted by the number of connections to Internet (hosts per 1000 people). Armenia is shown in a separate line.

The following figures characterize the information infrastructure of Armenia and allow us to compare Armenia with other countries.

Number of PCs in Armenia	**20,000** (my estimation)
Number of computers sold in Armenia or delivered by foreign aid annually	**4,000** (my estimation, taking into consideration government statistical data)
Number of Internet hosts in Armenia	**3,500** (my estimation, based on RIPE hostcount [2] for AM domain, consideration of NET and COM domains in Armenia and number of hosts behind firewalls)
Number of people online	**10,000** (my estimation)
Number of households online	**2000** (my estimation)
GDP per capita	**$430** (UNDP Armenia)

According to independent experts, the GNP of Armenia is estimated as US $3 billions. The estimation of the market value of Armenia is US $1.5 billions. And the Annual

Table 1

No	Country	GDP per capita (US$)	Market value (US$ bil)	Computers per capita (per 1000 people)	Connections to Internet (hosts per 1000 people)	Population (min)
1	United States	30,507	11,387	450	35.2	265.0
2	Sweden	25,559	332	353	29.4	8.9
3	Canada	20,391	611	364	29.0	30.3
4	Australia	21,274	292	366	28.8	18.5
5	Switzerland	35,421	687	299	22.6	7.1
6	Denmark	30,927	103	349	22.1	5.1
7	Netherlands	22,720	593	292	19.8	15.6
8	Singapore	31,167	83	316	13.3	3.1
9	UK	21,952	2313	283	11.9	58.7
10	Germany	25,363	1063	231	9.5	82.2
11	Ireland	19,079	64	263	8.4	3.6
12	Japan	33,655	2108	228	6.3	124.4
13	Israel	16,738	48	219	6.0	5.9
14	France	23,691	915	234	4.8	58.3
15	Czech Rep	4932	14	81	4.4	10.3
16	Hungary	4426	14	90	3.1	10.2
17	Spain	13,282	407	127	3.0	39.3
18	Italy	19,821	490	158	2.9	57.2
19	South Africa	3006	264	31	2.8	42.8
20	Portugal	10,079	67	103	2.7	10.0
21	Greece	11,423	53	73	2.5	10.4
22	Taiwan	13,119	275	147	1.7	21.6
23	Korea	9571	47	124	1.6	46.2
24	Poland	3516	13	55	1.5	38.7
25	Chile	4993	60	45	1.3	14.6
26	Malaysia	4724	85	65	1.2	21.3
27	Russia	3119	49	31	0.5	147.9
28	Mexico	4102	124	41	0.3	98.2
29	Thailand	3068	24	28	0.2	60.6
30	Turkey	3048	52	20	0.2	62.4
31	India	324	129	3	0.0	977.9
	Armenia	**430**	**1.5**	**7**	**1.1**	**3,1**

Income per capita can be estimated as 1.5 bil/3.1 $M = \$480$.

As Table 2 shows, Armenia has a very low GDP per capita [3] and market value. But due to grants, donations and assistance from different foundations, Armenia's place in the list does not strictly correlate with the GDP per capita, and the position of Armenia does not seems bad. This is mainly due to grants from Eurasia Foundation, NATO, OSI, AGBU, USIA, IREX, TACIS, and other organizations. The role of Eurasia Foundation should be specially underlined. Due to the policy developed by Eurasia Foundation, every grant includes hardware and software, as well as coverage of the communication expenses for the Internet connection.

One can easily note the correspondence between computers per capita and connec-

Table 2
Local prices for dial-up Internet access

Service provider	Dial-up e-mail	On-line e-mail	On-line Internet
Freenet (freenet.am)	free	free	
AUA (hragir.aua.am)	no	free	
Arminco	Reg. fee \$100 + 50\$/mth (80 Mb quota) + \$0.48 traffic fee		
ACC	Reg.fee \$50 + \$5/mth		Time limit daily 2.5 hrs (1hr working day + 1.5 hr evening): Reg. fee \$100 + \$10/mth Unlimited: Reg. fee \$200 + \$0.5/hr
Infocom			Reg.fee \$40 + \$50/mth
Byte			\$40 unlimited
Best			\$35 unlimited

tions to Internet in developed countries. However, connections to Internet in Armenia seem unusually high compared with computers per capita. This poinst to good starting conditions for the development of the Internet infrastructure.

There are other important parameters characterizing the technological progress of a country, such as percent of people online, cost to get online, credit/debit cards per 100. They are important for e-commerce, venture enterprises, and attraction of investments. Below, the Forbes magazine is quoted (the text in italics) because this analysis is very important for understanding the situation in Europe and interpolating the results for Armenia.

2. Situation in other contries

A recent report from Forbes Magazine [4] ranked the Internet markets of EU member countries, as well as US. The magazine ranked Internet markets in the EU on an A-to-D scale, ranging from excellent to unsatisfactory respectively.

It is interesting to cite the descriptions of problems in these countries and find out what are the main factors contributing to the progress of a country or to its lagging.

UK: *Tony Blair's* government has committed itself to one of the world's most ambitious Internet policies, aiming to get every British school child connected to the Internet by 2003. Moreover, the UK telecommunications market is the most liberalized in Europe, *with several upstart phone companies fighting to grab some of incumbent British Telecom's piece of the pie. Apart from the US*, no country in the world is more accommodating to entrepreneurs who need capital to finance their upstart Internet companies. *In sharp contrast with the Continent, floating stocks is a common means of raising capital. The UK market for venture capital is also thriving*. Britons have more debit and credit cards than any other Europeans, the importance of which cannot be overestimated for anyone who wants to set up shop on the Internet. *And finally, the UK has one enormous advantage that simply can't be overlooked: its language, English, which is understood not only in much of Europe but throughout the world.*

Sweden: the computers are new and the telecommunications infrastructure is of an exceptionally high quality. *Moreover*, the cost of getting on the Internet is among the lowest in Europe. *No wonder, then, that* there are more people online in Sweden than in any other country in Europe except neighboring Finland. *Because of the density of mobile phones in Scandinavia*, Sweden and its Nordic neighbors have pioneered the market for wireless e-commerce. *One example: As of April, users of some Ericsson phones in next-door Norway* will be able to buy movie tickets using their mobile phones as payment terminals.

Finland: Finland has the highest mobile phone subscription rate in the world. ... Finland has the highest Internet penetration in Europe. ... the most important factor is the country's low telecommunications access charges.

Denmark: strong technology infrastructure and stable economy. On average, the Danes have more computers than anyone else in Europe and plenty of phone lines to keep them connected. Twenty hours on the Internet costs only about $32 in phone and access charges, even less than in the US. As many as 13% of Danes have access to the Internet, more than twice the European average.

Netherlands: about 9% of the Dutch are connected to the Internet. ... The Dutch have a remarkably good technology infrastructure. Cell phones are everywhere ... It's no wonder that so many Internet companies are setting up shop in Amsterdam.

Norway: with so much available Internet access, one of the reasons why Norwegians are some of the most enthusiastic net surfers in Europe. Norway's GDP per capita may well be one of the highest in the world.

Switzerland: owing simply to the country's wealth, it remains one of Europe's most wired countries. Switzerland's wealth will probably attract Internet entrepreneurs in the years to come. ... On the other hand, Switzerland's hefty phone charges — the highest in all of Western Europe — are likely to discourage Internet development, ...

Germany: Germany's Internet access charges remain prohibitively high. ... If you want to hook up to the Internet via television cables instead, good luck: Deutsche Telekom is also the country's chief cable company.

Ireland: *Ireland's technology infrastructure isn't as advanced as Scandinavia's or even the UK's, and the country's wealth still doesn't measure up to its Northern European neighbors. So what makes Ireland one of the most attractive markets for Internet companies? The answer lies in the country's amazing economic recovery* — its GDP grew 47% over the past five years, *and its jobless rate dropped from 18% to 8% over the past decade — coupled with the government's consistently high tech friendly policies.* ... Ireland has aggressively courted foreign investments, especially in the high tech area. *It has worked very well: Today there are 1050 companies supported by the IDA in Ireland, employing 98,000 people. Among them: IBM, Microsoft, Oracle, Dell, Gateway and EMC.* The Irish telecommunications market was completely deregulated last year, *13 months ahead of schedule. What's more*, Ireland's telephone monopoly invested over $4 billion in upgrades and advancements to Ireland's communications infrastructure over the past decade ... With its English-speaking and well-educated, but relatively cheap, labor force, *Ireland offers a low-cost gateway to the UK, Europe's hottest e-commerce market.*

Belgium: Belgacom, the country's main telephone operator and Internet service provider, just announced cuts of 7% to 18% in Internet access tariffs.

Italy: telecommunications and ISP rates are among the cheapest in Europe, but cheap rates won't help as long as the Italians don't have computers to surf the web with.

There's a more profound reason behind Italy's lag: a lack of an aggressive, coherent government IT policy. *Compare Italy and Ireland, two countries at approximately the same level of economic development.* Ireland is courting foreign investors, especially tech companies, offering tax incentives and business development support. As a result, Ireland's technology and telecommunications infrastructure has improved dramatically over the past few years. *In contrast, Italy's inertia has produced one of the lowest Internet penetration levels in all of Western Europe.*

Spain: fighting against high telephone and Internet access charges, the Spanish Internet users launched the world's first strike in cyberspace *last fall by going offline for a day. Their rebellion quickly spread to other countries, such as France and Germany, where the simple act of getting online still gets the wallets of the web users.* At the end, the strike itself didn't have any impact on the telecom monopoly by Telefonica. But Telefonica and its troubled Internet service, Info Via Plus, got a lot of bad publicity. The strike forced the phone company to propose lower rates for the first time, and the Spanish government is considering further rate cuts. Technology infrastructure is among the poorest in Europe. There aren't many computers to go around, and phone penetration is also lower than in Northern Europe.

Greece: Greece is an Internet wasteland where only a madman would contemplate setting up an online business. One thing is certain: No country in Western Europe — not even Spain or Portugal — is further behind in Internet development than Greece. The country's technology infrastructure is years, maybe even a decade, behind the more developed countries in Northern Europe. Apart from the largest companies, hardly anyone owns computers in Greece. *Telephone penetration is much better — better, in fact, than the rest of Southern Europe and Ireland — but that doesn't help much with the shortage of computing power. What's more, getting online is very expensive.* In a country with one of the lowest disposable average incomes in Europe, it costs more than $60 to get online for just 20 hours a month. Needless to say, that's an almost insurmountable barrier to Internet development. It's hard to measure exactly how many Greeks are online today, but the best estimates are about 1%, the lowest figure in all of Western Europe. Greece's voice telephony monopoly, the OTE, has a firm grip on the market despite government plans to open it up for increased competition. Under the current plans, the OTE will keep its monopoly on fixed lines until 2001, exacerbating Greece's position as a laggard in Europe.

The results of the investigation are shown in the Table 3.

The estimation for the cost of getting on-line for private users in Armenia was done by survey of prices of local service providers. They are shown in Table 2.

Analyzing the report of the Forbes magazine it is possible to extract the main factors contributing to the fast development of a country.

- Modern telecommunications infrastructure;
- Deregulated economy;
- Accommodation of the country to entrepreneurs who need capital to finance their upstart Internet companies;
- Courting foreign investments, especially in the high tech area;
- Living standards (GDP per capita);
- Low Internet access charges;
- Low mobile phone charges;

Table 3

Country/region	Percent of people online	PCs per 1000	Cost to get online ($)	Annual disposable income ($)	Credit/debit cards per 1000
US	**16.0**	**580**	**34.87**	**21,928**	**148**
Europe Average	**6.0**	**352**	**49.32**	**14,801**	**39.31**
Sweden	15.0	509	36.6	13,457	44
Finland	14.6	505	19.77	12,309	23
Norway	14.3	515	35.36	16,966	44
Denmark	13.1	510	31.73	16,364	26
Switzerland	10.6	444	66.98	24,519	34
UK	9.5	441	46.17	15,996	100
Netherlands	9.0	450	40.23	14,062	21
Germany	6.4	362	68.43	16,996	17
Belgium	4.7	405	46.59	17,309	24
Austria	4.5	346	64.02	15,838	15
Ireland	4.0	404	50.57	12,107	26
France	2.6	368	48.28	15,685	39
Italy	2.4	297	35.65	14,266	14
Spain	2.2	202	42.09	9279	58
Portugal	1.9	160	46.12	7164	53
Greece	1.1	125	60.16	10,191	11
Armenia	**0.3**	**7**	**35.00**	**480**	**0**

- High density of phones and mobile phones;
- Liberalized telecommunications market, competition;
- Willingness of telephone monopoly to invest money in upgrades and advancements of telecommunications;
- Saturation of the country with PCs;
- Commitment of the government to national Internet projects;
- Considerable amount of debit/credit cards per capita;
- Presence of English-speaking and well-educated, but relatively cheap, labor force.

Unfortunately these factors are absent in Armenia. The monopoly on telecommunications of one company can bring Armenia to the end of list of world countries. The cost of getting online is too much for the countries with such a low GDP per capita. There is no government commitment to long-term Internet development plans.

Phone lines in Armenia are of bad quality. There are no expectations of considerable GDP increase.

What are the factors that may contribute to the development of Internet in Armenia?

Foreign aid, deregulation of the economy and the telecommunications market, government policy for attracting foreign investors, government control of phone and Internet access prices and special education programs.

How to increase the number of people online? The only way is to open public access sites with free access. There was a very good IATP-IREX project of this kind. Just to show its importance, here is some statistical data regarding IATP activity in Armenia.

AUA — 900 users (since April 3, 1998). The site is equipped with four computers and connected via Arminco.

NLA — 718 users (since March 24, 1998), 5 computers, via YerPhI.

YerIAC — 248 (since May 27, 1998), 4 computers, via YerPhI.

Gyumri — 236 (since July 11, 1998), 4 computers, via Shirak Internet Users Society (Arminco).

Vanadzor — 47 (since September 30, 1998), 4 computers, via Vanadzor Business Center (Arminco).

UNDP is providing training and free access of participants, including schoolchildren, to Internet.

One of the possible solutions for Armenia is the provision of free e-mail. Freenet is servicing 3000 e-mail users, and hragir.aua.am — 2350.

One of the other parameters describing the situation in a country is the number of households on-line. This points directly to the welfare and potential market of the country.

Larry F. Gladieux, a College Board researcher, notes that "*computers are in 75% of households with incomes over $75,000, but just under 20% of households making less than $15,000 have computers or daily access. Households on-line is one of the very useful parameters, characterizing a country as a potential market for e-commerce and investments. What are the characteristics of the worlds leading countries? Here are figures: two-thirds of US households will have Internet access by 2003*." The report estimates that US consumers will spend USD56 billion on Internet access services over the next five years, growing at a rate of 21 percent per annum. An estimated 45 percent of US households currently have a PC. 25 percent of US homes are estimated to have Net access, and this is expected to increase to one third of all households by the end of the year. In contrast, *Yankee estimates* that only 5 percent of Western European homes currently have Net access, increasing to 17 percent by 2002. A survey indicates that there are 960.000 households, i.e., one in 25 British households, in Britain who have access to the Internet, a figure which has more than doubled in the last twelve months. In Finland, home access has taken the lead as the most popular place from which to access the Internet. The Internet is the primary use of the computer for 46 percent of US households, followed by home office use at 23 percent and games and entertainment at 18 percent.

The number of households on-line directly corresponds to the quality of life. This figure will remain very low in Armenia, since there is nothing to indicate a sharp increase of the living standard in Armenia.

If we compare Armenia with neighbouring countries we see that Armenia is a bit ahead of Georgia and Azerbaijian. This is due to liberal registration rules in AM domain. The Estonia advance in Internet is amazing. The same post-Soviet country managed to go far ahead of all others. Israel is shown as an example of a very advanced country.

Now the analysis of different aspects of Internet usage follows.

3. Education

The following figures characterize distance education. At present 750,000 students are involved in 26,000 online courses. Percentage of US public schools connected to the Internet, in 1996: 65.

Table 4

Country	SOA	Number of hosts
AM	315	1210
AZ	117	531
GE	76	960
EE	3463	28031
IL	12280	141584

The US Department of Education says that, largely thanks to government subsidies, 51% of classrooms, school computer and science labs, and school libraries had Internet connections in the fall of 1998 (compared to 27% in 1997 and only 3% in 1994). Smaller and poorer schools are now just as likely to have Internet connections as larger and wealthier schools. (Reuters/San Jose Mercury News 1 March 99.)

2.4 million UK children between the ages of 6 and 16 have used the Internet, according to a study by NOP Research. Forty three percent of British schools are connected to the Internet, a BESA report has found.

Distance education can be introduced in the higher education institutions in Armenia. Most of the universities in Armenia are computerized and Internet connected. They are involved in different projects and gradually improving their infrastructure. The problem is the necessary bandwidth.

The situation in Armenia with respect to school connectivity to the Internet is disappointing. The government seems does not seem to care about that. There are efforts by some organizations, like Narod, to bring e-mail and Internet to schools.

The Narod Network Project links schools around the world in a collaborative network on the Internet. Through specially designed activities, students learn to use information and communications technologies to become acquainted with their global peers, to do research and to publish their work online.

The first Narod Network Project, NNP'98, involved 10 schools in five countries. NNP'99 had more than 30 participating schools in eight countries. NNP'00 will reach 15 countries and includes more than 60 schools, 25 of them in Armenia.

The Eurasia Foundation had given a grant for setting up an Internet training center in Children Library. Without the government involvement, little will change in Armenia.

Armenia needs the "Internet to every school" project.

Getting schools on the Internet is one of the primary tasks of the nation entering the Information Age. Children connected to the Internet will be able to talk directly to schoolchildren all over the world, explore rich storehouse of knowledge at museums, libraries, and universities. Internet access will make schoolchildren computer literate and better prepared to work in the technology intensive jobs of the future. The project will play an important role in preparing young people. The connectivity to the Internet from a networked classroom will be one of the great teaching aids of the next century. The project must include the following components: computer, network, software, Internet access, and training teachers and students.

Because of small budgets the technology offered must begin a very modest way — with a single computer and modem — and then expands to complex LANs.

Dealing with network administration and technical issues presents another area of potential difficulty.

The projected starting price of an Internet capable system is $1000. This includes a computer, modem, dial-up connection to ISP and the Internet connection software. Schools must be connected to the Armenian Scientific and Educational network. The ISPs should be reimbursed by the government to cover additional expenses and servicing school connections. One of the important aspects of the project is to increase the skill of the children in independent learning, so that they are able to look for information on the Net.

There is also a great need for training Internet specialists in the country. Just a few examples:

France has recently revealed that it will spend USD 8.3 million on training managers for the Web this year.

In Canada, political support for Internet initiatives is strong and the federal government is driving an ambitious agenda to make Canada the most wired country in the world by the second millennium. They announced a budgetary investment of USD 1.2 billion in IT projects, incorporating a USD 134 million investment in the education system and in community access programs. The initial objective of 'SchoolNet' was to provide 16,000 schools and 3400 public libraries with Internet access by the end of March 1999 while the Community Access program aims to have 10,000 communities across Canada online within two years.

4. E-commerce

According to figures published by International Data Corporation, USD 8 billion worth of business was transacted on the Web in 1997 and that figure will grow to USD 333 billion by 2002.

Sixteen percent of Dutch users buy online. Dutch users are expected to spend a total of USD 44.7 million on online purchases by the end of 1998, according to Blauw New Media.

In Western Europe, e-commerce generated USD 5.6 billion in 1998, with the vast majority of revenue earned in business-to-business transactions, according to IDC Research. UK consumers will spend over USD 9.8 billion on the purchase of goods online in the next five years, according to Verdict Research.

E-commerce revenue in Europe will surge to over USD 8 billion by 2004, up from USD 35.8 million in 1997, according to a report by Frost & Sullivan.

The Internet is the primary use of the computer for 46 percent of US households, followed by home office use at 23 percent and games and entertainment at 18 percent.

Is Armenia ready for the digital economy age? Definitely not. There is no local Intranet. There is no enough presence of companies on the Internet. There are no debit/credit cards in the hands of population. The very low GDP per capita points to the low buying capacity of the population.

Still, we must not forget that the Internet is the World and if any local company has something to sell the whole world is the market. The company must invest some money in the Internet presence, provide the acceptance of payments through debit/credit cards, and quick and safe delivery of products to customers. In this way Armenia can be involved in world e-commerce.

5. Communications

Several messages are listed below just to illustrate the situation in other European countries.

First Pan-European Net strike

Feb 01 1999: Internet users in Belgium, France, Italy, Poland, Portugal, Spain and Switzerland were asked to boycott the Net for 24 hours yesterday, in protest to high access charges.

Second Internet strike planned in France

Jan 14 1999: Keeping pressure on the national telecommunications provider to reduce the cost of Internet access, French users are planning another strike for January 31, 1999.

Czech Netizens go on strike

Nov 11 1998: In what is rapidly becoming the standard means of protest against high telecommunication charges, Czech users and Web operators are planning an Internet strike.

Germany goes on Internet strike

Nov 02 1998: Over 6000 German Internet sites went off-line yesterday to demonstrate against the cost of Internet access in the country.

It sounds strange but most of European countries are facing the same problems as Armenia. They are Telcos monopolies and the high prices of a connection.

The Forbes magazine exhorts Europeans to get rid of their ancient telecom monopolies with their absurd telephone charges, take a cue from the UK and pave the way for flat rate charges.

Though there is no way to overcome the monopoly and high prices of local Telcos, downlink and DirectPC technologies can considerably save ISPs and users expenses.

The connectivity of Armenia to the Internet can be characterized by the following figures:

ISP	Two-way channel	Downlink channel
Arminco	Teleglobe, 512 Kb	Interpacket 10 Mbps
Yerphi	DESY, Germany, 192/128 Kb	DirectPC, 14 Mbps
	Teleglobe, 64 Kb	
WEB	Teleglobe, 64 Kb	
Infocom	Teleglobe, 256 Kb	
ACC	Teleglobe, 256 Kb	
MEGACOM	Other, 64 Kb	
Freenet	Holland, 16 Kb	
Total	1.5 Mb	

Most of the ISPs networks are radionetworks. Yerphi, Web has 60 radiobridge access points, Arminco — 40, NAS —12, Infocom — 4.

All Teleglobe channels are provided by the local Telco Armentel. The prices are:

Capacity, kb	Price in USD/month
64	3000 + 20% VAT
128	5600 + 20% VAT
256	10,800 + 20% VAT
512	19,200 + 20% VAT

The prices are high and we can hardly expect fast increase of Internet usage in Armenia. The overall Internet connection capacity of Armentel is 2 Gb, but because of high prices this capacity is not used.

There is a way of remedying this situation: the development of academic, research, school non-commercial networks with government support. Non-commercial Internet connections are not subject to the limitations caused by Armentel's monopoly and can be set up by contracts with international ISPs at low prices.

The rapid advance of the Internet in Armenia and the arrival of many Web servers started to create more and more local content and the importance of powerful intra-Armenian communication has increased.

6. Internet presence

The number of w3 servers is one of the basic characteristics of a country's informational infrastructure. There are more than 110 Web servers in Armenia now and their number is increasing.

References

[1] http://www.nua.ie/surveys/analysis/graphs_charts/ 1998graphs/primaryuses.html.

[2] RIPE DNS Hostcount by Mirjam Kuehne, [mirjam.kuehne@ripe.net].

[3] Human Development Report, Armenia, 1998, the role of the state, UNDP.

[4] http://www.forbes.com/tool/toolbox/billnew/ countries.

[5] Nua Internet Surveys: April 6, 1999.

Jacek Gajewski

CEENet Educational Activities

One of the main activities of CEENet is the education of young networkers and managers. Several discussions during the General Assembly meetings of CEENet have to the conclusion that education is a major need in our region, and on a larger scale it is the right investment to diminish the gap between CEE and the Western world.

There are several types of educational activities, which are carried out (or supported/planned) by CEENet: training workshops, distance learning courses, vendor constructed courses and SONETA training facility.

1. Training workshops (ANW) are our main activity (this topic will also be discussed in the following talk by prof. O.B. Popov) — they are of two different types: managerial methods and network technology.

a. Managerial methods workshops are organized biannually: there was one in Yaroslavl (Russia) in 1998 and the next one is planned in Ohrid (Macedonia) in June 2000. The main motivation of the managerial methods workshop is that the new emerging NRENs need their managers to be trained in various non-technical topics (e.g., how to deal with PTTs, how to apply to local and international funding bodies, how to construct agreements on peering and Internet exchanges, what should be the domain name policy, etc.).

These workshops usually gather 50–60 persons from $\sim$ 25 countries, they last $\sim$ 3 days and the cost is 60–70 k$.

b. Network technology workshops are organized annually: Warsaw'95, Budapest'96, Zagreb'97, Bratislava'98, Budapest'99 and again Budapest'2000. They are our response to the overall need for network experts in CEE region. Those workshops include 90+ students, from 23 countries, 15+ lecturers and 6 persons of staff or instructors. They last 10 days and cost 150–170 k$. We try to attract vendors and sponsors, e.g., last year we received a software worth 10 k$ and 600 O'Reilly books worth 35 k$ as a donation (the books were distributed to all students).

The 2000 workshop will be proceeded by three distance learning (DL) courses, which will give students the basic knowledge — in order to the entry level, which might be very different among students from different countries. Out of the $\sim$ 300 students who will take DL courses, around 100+ will be selected to continue in a face-to-face mode in Budapest 2000. The DL courses will be run by 9 instructors, will last 3–4 weeks, with the average load of 10 h/week. They cost is estimated to about 40 k$, but this will be definitely less in next years, when all the courses finally set up.

2. Distance learning (DL) courses are at the moment used for the preparation and the selection of the students during further face-to-face training.

In principle these same courses could also be used for a massive training. We offer DL courses on DL methods itself as well as on different aspects of networking (technology, content creation, user services, etc.).

3. Vendor constructed courses have sometimes a very high educational value and lead often to certification. We are closely cooperating with Cisco on setting up several "Cisco Academies" in CEE region. At the moment we are investigating educational initiatives of other corporations (Sun, Oracle, Microsoft, 3Com). Prior to the Managerial Methods ANW in Ohrid we plan to organize an "Educational Day", where corporations can present their educational programs.

4. SONETA is a planned semi-permanent networking training facility at the Central European University (CEU) Conference Center in Budapest. Some investments (in this already very well computer and network-equipped center) have already been made. About 30 k$ will be invested to enhance the teaching facilities. The CEU runs also the CISCO Academy — which gives additional possibilities for training in routing technologies. CEENet will use this facility only during a small fraction of the year — in our opinion it could and should be used by others and we will welcome any suggestions for cooperation.

Oliver B. Popov

CEENet Comprehensive Educational Endeavors in Networking

1. Introduction

The Central and Eastern European Networking Association (CEENet), a legal entity registered under Austrian law, is an association of national organizations whose primary mission is the international co-ordination of the establishment and operation of academic, research and education networks in the CEE region and adjacent countries. The current membership of the organization includes twenty-one national research and education networks from the following countries: Albania, Armenia, Austria, Bulgaria, Belorus, Croatia, Czech Republic, Estonia, Georgia, Hungary, Latvia, Lithuania, Macedonia, Moldavia, Mongolia, Poland, Romania, Russia, Slovenia, Slovakia and Uzbekistan [9, pp. 31–39].

One of the first activities of CEENet was to embark on the road of education, since it is apparent that the widespread usage and acceptance of the Internet and networking technology in general depends on well-trained and educated individuals who be involved in building the infrastructure, the services, the administration and networks in their respective countries. Hence, the Association, based on the previous experience by ACONet and the ISOC workshops on networking for the developing countries, decided to establish a series of educational workshops [3]. Throughout the years, the institution of workshops has proved to be a powerful vehicle to introduce and promote new paradigms related to the areas of network technology, network management and policies [9, pp. 31–39].

2. The engineering and the content go the distance

At all CEENet educational activities, the attendees learn through lectures, labs, roundtable discussions, and group assignments advanced networking concepts, to meet the network leaders from the West including the representatives from the CIT companies, and to identify possible venues of co-operation, common problems, and possible/acceptable solutions.

The generous awards and grants by NATO Science Division and the Soros Foundation (Open Society Institute) provided a sound financial basis for the education and training of more than 500 individuals from CEE countries. The support and assistance of many world-renowned vendors and corporations of computing and communication technology such as Cisco, Sun, GlobeComm (NetSat Express), 3Com, Silicon Graphics, Newbridge Networks, Hewlett Packard, Oracle, and O'Reilly enabled the attendees to experience state-of-the-art equipment and laboratories. Since 1995, there have been five network technology workshops in Warsaw, Budapest, Zagreb and Bratislava, and again in Budapest [5].

The international team of CEENet lecturers, consisting of distinguished scholars and experts in network technology and its applications, covered an extensive set of topics into two parallel tracks, Engineering the Network and Network Knowledge Systems, respectively. The curricula included, but was not limited to the following subjects: network addressing and topology, DNS, authentication and authorization, routing, congestion and flow control, dial-up networking, virtual private networks, security, ATM, management of a network infrastructure, Web technology, the structure and the architecture of network and Internet information systems and data-bases, e-commerce, cashing, mirroring and replication [6].

In 1998, a third track was added. It became an integral part of the workshops, and addresses the issues of distance and open learning. In the true spirit of networking and the Internet, during the 1999 workshop, there was a pre-workshop distance learning course on the basics of distance education and course design. For the Network technology workshop planned for 2000, the design of pre-workshop courses for the engineering and content tracks is already underway. The pre-workshop courses have, among others, two main purposes: to assist in the selection process of the best possible candidates, and to provide them with a fundamental knowledge in one of the tracks, so that the actual attendance at the workshop is devoted to more advanced topics.

The workshop returned to the Budapest venue in 1999, at the CEU Conference Center. The possibility of using this outstanding facility is one of the reasons behind the idea of SONETA, or the School of Networking Technology and Applications, a more permanent institute for the execution of the CEENet education efforts.

3. Policy and management do make the difference

The necessity to address, and if possible to define, a common set of problems and their resolutions beyond the technology level (organization, administration, funding, and cooperation) have induced the need for another type of a workshop. Running a National Research and Education Network requires not only networking technology skills, but also a deep understanding of managerial and policy issues. The need for this type of workshops is also a recognition that telematics or communication and information technology, is the precondition for the economic growth and general progress of each society. Due to the complexity and the globality of the effort, there is a need for policy recommendations and implementations on each level, so that public and social awareness creates the necessity of realization. So far, there have been two workshops on network management and policies, the first one in Tartu and the second one, done jointly with TERENA, in Yaroslavl.

The First CEENet Workshop on Network Policy was held in May 1997, in Tartu, Estonia. There were sixty-three participants, from twenty-three countries, who presented forty-four papers. The profile of the participants included directors of the CEE NRENs, senior representatives from governments (such as ministries and agencies responsible for national research and education networking), and delegates from institutions of higher education. The objective was to confront the ideas of these different players in the research and education networking, namely to identify major problems and issues, to recognize the main characteristics of NRENs, to enumerate the list of goals and priorities, as well as to define a set of strategies, and finally to provide a common set of guidelines useful for all the NRENs in their respective countries [6, pp. 9–14].

Three major topics were discussed in Tartu:

- Network organization and administration (infrastructure, administrative entities, fundamental activities and departments, public relations, education, and research);
- Financing and funding (funding models, major sources, international membership fees, types of traffic, charging, profit vs. non-profit approach);
- Networks in National and International Context (relations with other operators and providers such as PNOs, Telecoms, and other commercials, peering agreements, national co-operating agencies and institutions), international co-operation and projects, and the role of the national network in building the information society).

The result of the conference in Estonia, was the CEENet Tartu Declaration which is a series of statements on the status of research and education networking in each member country, and a set of recommendations to the governments, the national networks, international agencies and organizations.

There are seven parts in the Declaration: the preamble, the challenges, the current situation, the goals and the objectives to be reached, the means of achieving the objectives, the list of problems, and a set of recommendations for the governments, the national networks, and the international organizations and agencies. One of the recommendations to the governments is that they must acknowledge the inevitability of science, education and technology, and thus provide a fruitful and stimulating environment by enacting precise and non-monopolistic telecommunication laws, protect intellectual rights and freedoms, as well as the security and the integrity of the data. On the other side, national networks should observe international standards with respect to the infrastructure, the quality of the services, and the right to access. An intensive effort should be allocated to primary and secondary education since they play an important and lasting role in the dissemination of the Internet and network/information literacy. While the Declaration is an aggregation of recommendations, rather a binding resolutions, it profoundly and strongly defines the objectives and the principles how to converge and integrate into the Global Information Society.

The next workshop on the management issues in running a national network was done jointly with TERENA in Yaroslavl, Russian Federation. Now, the top officials responsible for running the national networks in the CEENet member countries became "students", while the lecturers were the managers and the directors of the most advanced research and academic networks in Europe, TERENA and DANTE officials.

The objectives of the Yaroslavl Workshop were to identify the structures required to provide networking services to the research and education community, to define the principles that should be observed in order to achieve and maintain utility of services, technical excellence and financial stability, and to understand the funding mechanisms and the possible ways of their implementation, naturally according to the characteristics of the country and its policies [9, pp. 15–20].

The first two sessions of the Workshop consisted of the TERENA and DANTE presentations with respect to their role in the European research and education networking, and the enumeration of the Workshop programme and the explication of the expectations of the participants. The workshop continued with an excellent statement of what a science policy is, and an overview of its attributes relative to networking. Then the identification of new user communities and groups followed, which usually falls in the realm of research and education networking, such as schools, museums, municipal centers, and libraries.

The fourth session dealt with the basic services that NREN network should offer: connectivity, services, licenses, discounts, training, education, research and development. The very important issues with respect to the telecommunication laws and regulations, the emerging technologies and how they affect the communication market, were the focus of the next two sessions. The last four sessions were devoted to the management of the NREN, illustrated with case studies, as well as the relations with the government and funding mechanisms.

4. Conclusions

In order to document its activities and disseminate its results and accomplishments in an efficient and lasting manner, CEENet has established a series of publications. The books from the CEENet Edition, which primarily contain the material presented and lectured at the workshops and the conferences, are distributed to the member countries, corporations, international organizations and governments.

The ability to learn is one of the primary characteristics of human civilization. Moreover, the capacity to organize and systematize what has been learned, and to convey the gathered knowledge, both empirical and theoretical, through the process of teaching has been one of the principal factors of human progress. The past activities of CEENet are the tribute to these crucial human activities, which have new qualities and complexities in the Information age, so eloquently and strongly expressed by the NRENs [6,7,9]. Hence, CEENet in the years to come, and in particular in 2000, plans to continue with its education efforts through a managerial networking workshop in Ohrid, Macedonia and the network technology workshop in Budapest, Hungary.

These events reflect the temper of our times and represent the pioneering effort of CEENet to implement the idea that networking is for everyone everywhere. The CEENet educational endeavors have had unique and unqualified support by NATO, OSI and the NREN member organizations. These activities are probably one of the most gratifying investments in the future of the CEE countries.

References

[1] Gajewski Jacek, private communication, 1996.

[2] Hofmokl Tomasz, Academic Networks in Central and Eastern Europe, INTEROP'95, Frankfurt, 1995.

[3] Rastl Peter, Coordinating Networks in Central and Eastern Europe, CEENet, The Proceedings of INET'94/JENC 5, Prague, June, 1994, Book2, pp. 424/1-424/3.

[4] Vietsch Karel, private communication, 1998.

[5] Popov Oliver B. (Ed.), The Road to Global Connectivity, Proceedings of the First CEENet Workshop on Network Technology, CEENet Editions with NATO ANW Series, Warsaw, 1996, pp. 9–12.

[6] Popov Oliver B. (Ed.), National Networking Effort: Concept and Strategies in Visionary Ideas for a Visionary Future, Proceedings of the First CEENet Workshop on Network Policy, CEENet Editions with NATO ANW Series, Prague, 1998.

[7] http://www.ceenet.org

[8] http://www.terena.nl

[9] Popov Oliver B. (Ed.), The Avant-garde of the Information Society, Proceedings of the First CEENet /TERENA Workshop on Network Management, CEENet Editions with NATO ANW Series, Prague, 1999.

Networking Developments in the Caucasus Region
J.-P. Nadreau and O.B. Popov (Eds.)
IOS Press, 2001

Anne Villems

Distance Education — a Tool to Change Society

Where is Estonia?

Islands: Saaremaa and Hiiumaa

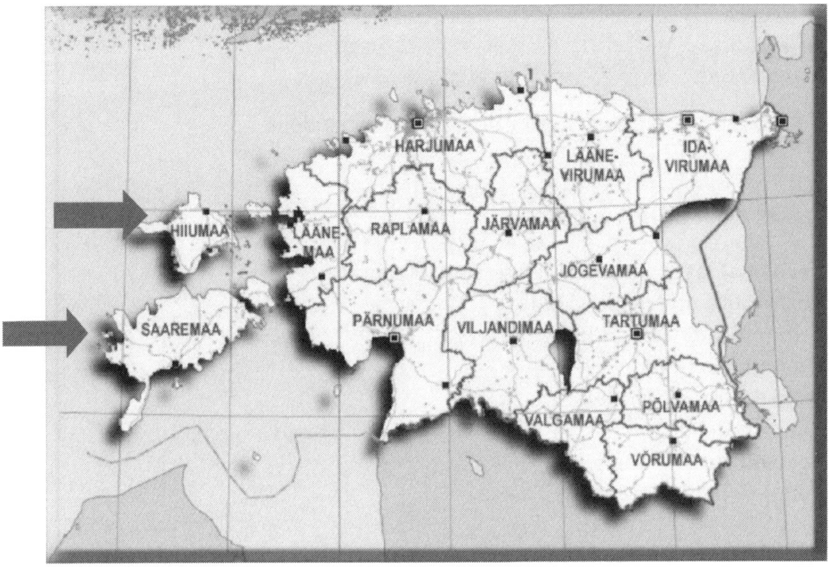

Differences

	Saaremaa	Hiiumaa
	30 000 inhabitants	3 500 inhabitants
	hard working closed, no jokes	open, joking, first Internet
	During the course	
	many new ideas, open, humorous . . .	no new ideas, close, no humor . . .

What happened? Why?

The only real difference:

Five years ago the University of Turku (Finland) opened its distance education center in Saaremaa and organised courses on a regular bases.

Conclusion:

Distance education can be a tool for making changes on a social level.

Distance education

Wide range of distance education (de) tools and means

- From elements to enrich the face-to-face (f2f) communication with e-mail, lists, www, video-conferencing, etc.

- Through combined settings:
  - f2f–de–f2f (classical),
  - de–f2f (T3 WirEd),
  - de–f2f–de;
- To full DE, where the teacher never meets a student.

What makes a good DE course?

- Technical level works with a technical support;
- Team work on teachers' side;
- National language use;
- Communication;
- Active learning model;
- Tutor training;
- Specialized software;
- People who believe in DE.

1. Properly working technical level

- Electricity, computers and Internet exist and are working;
- Technical support at hand when needed;
- DE has to stay on a proper level
  - mainly well known technical level,
  - can have advanced elements, but in limited amount.

2. Teamwork — on teachers' side

- To develop a good DE course takes huge amount of time:
  - good structured learning materials,
  - well designed activities and communication plans;
- All good experienced teachers are always very busy;
- Cooperation on teachers' side and tutors' is the solution.

3. National language

- For learners there are too many new things:
  - new subject,
  - new activities,
  - new attitudes,
  - new kind of tasks;
- If this all happens in a foreign language, you cannot deliver it to a significant number of students.

4. Communications

- Student failures are mostly due to lack of communication;
- Teacher–student communication
  - tutors will help, tasks have to be well designed;
- Student–student communication
  - specially suitable to adult learners (develop trust!),
  - examples: Peer reviews, informatics teachers case.

5. Active learning

- Materials on the Web — this is not a course!
- The whole learning process has to be prepared and an active role must be given to the students.
- Learning happens through doing.

6. Tutor training

- Tutors are helpers of the teacher/professor;
- Tutors have to understand their role;
- Tutors have to know the subject;
- Tutors have to be trained;
- Tutors are supporters, not selfish;
- Where to find good tutors? You have to train them!

7. Specialized software

- Materials on the web, group task support software, list/news group for communication, synchronous tools
  - you cannot manage it without special software;
- Many good systems exist — we chose WebCT — one of the most popular, good price politics
  - http://www.webct.com/.

8. People who believe in DE

- People who will set up the DE system have to believe in it
  - there are many possibilities to make it ineffective.

Long time needed

- From the hardware aspect view point: we need more computers / bandwidth / routers;

- From the “soft” aspect view point:
  DE environments, DE-support services, trained tutors, instructional design knowledge, curriculum development models, tutor training packages, etc.
- The latter needs are more difficult to fulfill, but the hardware has to work!

Both sides have to be ready

- The learners
  - their hardware and connection,
  - their motivation,
  - their ability to work with DE environment;
- The trainers
  - their hardware and connectivity,
  - their DE know-how,
  - course materials, they can use for tailoring.

Nicolai Sandu

Romanian Education Network

RoEduNet is an organization and a data communication system which connects a number of Local Area Networks through their Central Access Nodes to a national WAN data communication infrastructure, operated by NSP (Network Service Provider) nodes in Bucharest, Iasi, Tg. Mures, Cluj, Timisoara and Craiova. The aim of this technical complex is to offer the participants — universities, cultural and scientific nonprofit institutions — the means to communicate with each other as well as to have access to the Internet.

Who are We?

The Romanian Education Network (RoEduNet) was created in July 1993, once the installation of the P.U.B. Local Area Network was accomplished and the P.U.B. Central Node got connectivity with the international data networks and services. From the very beginning RoEduNet was conceived as an open structure, offering free access to academic, scientific and cultural non-profit institutions. Once the first institution was connected — the University of Bucharest, in August 1993 — the nucleus of the academic data communication infrastructure was created.

This process has continued and, today, after six years, the RoEduNet data communication infrastructure covers the national territory, connects and offers services to more than 150 institutions, offers international connectivity through two high speed channels — **1.5 Mbps with TaideNet and 4 Mbps with LoralOrion**. The measured traffic figures shows a data exchange volume over the international channels of 0.8–1 Tbytes/month. The structure remains open to all universities as well as to non-profit scientific and cultural institutions.

In the meantime, a growing process has taken place. This process followed a bottom-up strategy, namely each new node in the network offering access as well as assistance to any new institution required to be part of and to operate a node in the RoEduNet data communication network. The process was far from being anticipative, did not have a program, did not follow a project. It was rather adaptive, heterogeneous — the growing complexity of the topology not being followed by the necessary upgrade of the technical configuration of the nodes, by the creation of the administrative and operative structures, adequate to the new conditions. From this point of view, the RoEduNet data communication infrastructure reached a critical level of complexity.

Another significant aspect relies on the status of the RoEduNet nodes; none of them used to have a distinct functionality or status as WAN node within RoEduNet. All of the nodes were central to the Local Area Network of the institution to which they belong, with certain responsibilities within the RoEduNet data communication infrastructure.

Considering the above-mentioned facts and the possible consequences, the RoEduNet data network has been restructured, according to the hierarchical model. The status of the participating nodes now makes a clear distinction between NSP nodes (Network Services Provider) and End User, client nodes. NSP nodes and the links between them define the RoEduNet backbone, which is level 0 in our model. NSP nodes are also regional POPs, concentrating traffic from their service area. The end user nodes, level 1, connect to the nearest RoEduNet NSP node. The NSP nodes are located at the central nodes of several universities in large academic centers throughout the country: Bucharest, Iasi, Cluj, Targu Mures, Timisoara, Craiova. Even this model has been superseded, by changing the designation of level 1 and moving end users to level 2. Level 1 is now occupied by POPs which will be deployed in each county's capital city.

RoEduNet: development plans

As presented before, RoEduNet can hardly be described as an Internet Service Provider. A more appropriate definition would be: *RoEduNet is both an organization and in the same time a data communication system which connects a number of Local Area Networks through their Central Access Nodes, thereby creating a national WAN data communication infrastructure. The aim of this technical complex is to offer the participants — universities, cultural and scientific non-profit institutions — the means to communicate with each other as well as to have access to Internet.*

The RoEduNet data communication infrastructure is structured according to a hierarchical model.

Level 0 — backbone

RoEduNet Service Provider nodes, or NSPs. These nodes have the task of operating the RoEduNet backbone and providing connectivity to regional networks as well as End Users. The backbone links are presently running at 2 Mbps. NSP nodes are operated in 6 major cities: Bucharest, Iasi, Targu Mures, Cluj, Timisoara and Craiova.

Level 1 — POPs

RoEduNet Points of Presence. They get connectivity from level 0 nodes and provide network services in the capital city of each county.

Level 2 — users

RoEduNet "clients". They get connectivity from level 1 nodes. NSPs also act as POPs in their county.

Here are brief descriptions of our short-term development plans.

Upgrading the backbone and Internet access

This stage is complete.

All of the nine 2 Mbps links between NSPs are operational: Bucharest–Iasi, Bucharest–Targu Mures, Bucharest–Cluj, Bucharest–Timisoara, Bucharest–Craiova and Craiova–Timisoara, Iasi–Targu Mures, Targu Mures–Cluj and Cluj–Timisoara. The backbone is resilient in case of link failures, by the quick convergence of connections through the available channels. Load balancing across multiple paths is also being considered.

The existing 1.5 Mbps Internet link to Europe was supplemented by a US satellite connection, operating at 4 Mbps. We are envisaging an upgrade of the European link to 2 Mbps. This kind of connectivity, both in terms of speed and reliability because of the two redundant paths, will effectively be the best Internet connection available in Romania.

Deploying POPs

RoEduNet Points of Presence (POPs) will be deployed in each county's capital city. A 256 kbps link will connect the POP to the nearest NSP node on the backbone. The topology for regional access will be initially star-shaped around each NSP node. Backup lines for POPs will probably be added in the future. About ten POPs are in various stages of development.

The configuration of a POP will include a server running Linux, an access server and modems and a wireless access point for metropolitan access.

User services

A trouble-ticketing system has been tested and will be integrated in the network management system developed in-house. All events pertaining to the availability of our network (link or server failures, misconfigurations, etc.) will be logged using this system. An automated report will be generated and sent to the interested parties.

Web hosting will be available soon for all our users who do not want the responsibility of running their own web site. Also, this would help relieve the "pressure" on our communication channels.

Since the most important resource in networking is the human resource, we will organize training seminars on system administration and networking.

Network topology

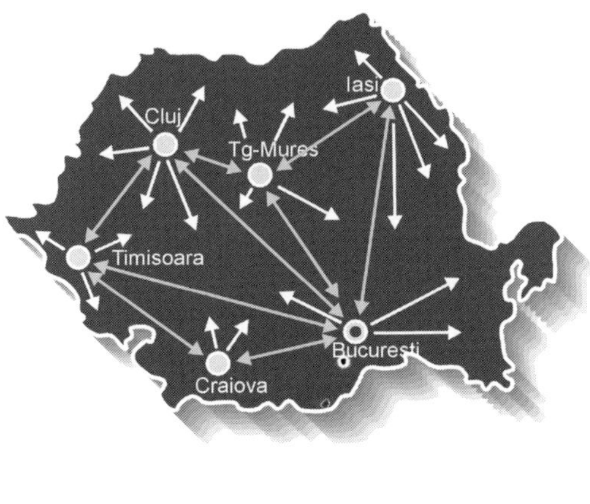

RoEduNet — backbone

The RoEduNet backbone consists of the links between the NSP (Network Service Provider) nodes in six major cities: Bucharest, Iasi, Targu Mures, Cluj, Timisoara and Craiova.

These nodes make the "level 0" entity of our network, the infrastructure (backbone) providers. The blue (2 Mbps) lines represent the existing links between the backbone nodes.

The green (256 kbps) lines are the connections between RoEduNet NSP nodes (backbone nodes) and RoEduNet POPs in each county capital city. These connections will bring a RoEduNet point of presence closer to educational institutions throughout the country, minimizing the cost of the network. The Internet connection is provided at the Bucharest node, using VSAT, to Taide Network and LoralOrion. Internet access capacity is 1.5/0.5 Mbps and 4/1 Mbps, respectively. We are planning an upgrade of the European link to 2/1 Mbps.

Dynamic routing is used to provide the desired redundancy, as well as to establish gateways with other networks in the NSP nodes. We have established gateways with RNC, KappaNet, DNT, Logic Telecom, Romania Data Systems and EUnet in Bucharest, Codec and DNT in Cluj, Iasi and Timisoara. A document describing the routing architecture on the RoEduNet backbone will be made publicly available. All NSP nodes are/will be able to connect our users via dial-up, leased asynchronous lines and leased digital (n*64 kbps) lines. Based on local resources, users can be also connected over Ethernets, cable (CATV), wireless or fiber optic. NSP nodes are sufficiently equipped for operating a high speed infrastructure based on this model, high-performance routers being deployed in each node.

Ongoing projects

Bucharest: Wireless Metropolitan Area Network

Considering the high cost and low availability of high-speed Telecom lines, even in Bucharest, a Wireless MAN will be developed to connect our users. It will provide a reliable 2 Mbps link at an approximate cost of $1000. It is the continuation of a pilot project, which already provides connectivity for about a dozen institutions.

Bucharest: Development of a Search Engine for the Romanian Web

The traffic patterns (in/out ratio) we see on our lines show that Romanian Internet users do not easily find the information they require on local servers. By searching only on Romanian servers this search engine will hit local servers, thus decreasing the load on the international lines. It should also encourage the development of content-rich Web-sites.

Cluj: University Network Development of a Network Monitoring Software System

This will be a Java-based Network Monitoring Tool helping us to effectively manage our large network. It is backed by a mSQL database and a Web server. The system allows the design and display of topological maps with information derived from SQL queries.

On a waiting list ...

Bucharest: upgrade of the "Politechnica" University campus LAN. Entirely based on 10Base5 Ethernet, the PUB LAN is now obsolete. We are looking at ATM and/or gigabit Ethernet to upgrade it.

All

Creating RoEduNet PoPs in every county (administrative district) of Romania.

For more information please contact http://www.roedu.net.

Networking Developments in the Caucasus Region
J.-P. Nadreau and O.B. Popov (Eds.)
IOS Press, 2001

Jacek Gajewski

Summary of the Discussion during the Parallel Session on Training

The session was attended by 18 persons. A formal presentation on "Cisco Academy" project by Nicolai Sandu was followed by lively discussion on training needs in the Caucasus region. It was common understanding that there is a need for: networking experts, network administrators, qualified instructors and network managers.

In many cases certified experts are needed, but usually vendor run courses are rather expensive. Funding bodies can help by forming "friendly" environment to vendor certified courses (e.g., cover travel and accommodation costs) but cannot pay directly for participation fees. An example is the initial training of the lecturers for the Regional Cisco Academies. Due to the special rules of the Cisco Academies the tuition is free but travel and accommodation has to be paid for 6 persons per country (round trip to Birmingham, total stay of 18 days). We received from OSI their support for the creation of local Cisco Academies in CEE countries. With some enterprising skill Academies might be self-sustainable, which opens possibility is for the cooperation with the World Bank.

Telemedicine training projects, before going full scale, should prove their feasibility first on a smaller scale in the Caucasus region.

The important issue of training scientists (other than in computer science, e.g., from humanities, older ones) has been mentioned.

Author Index